CALCULATION EXAMPLES

ARITHMETIC 3

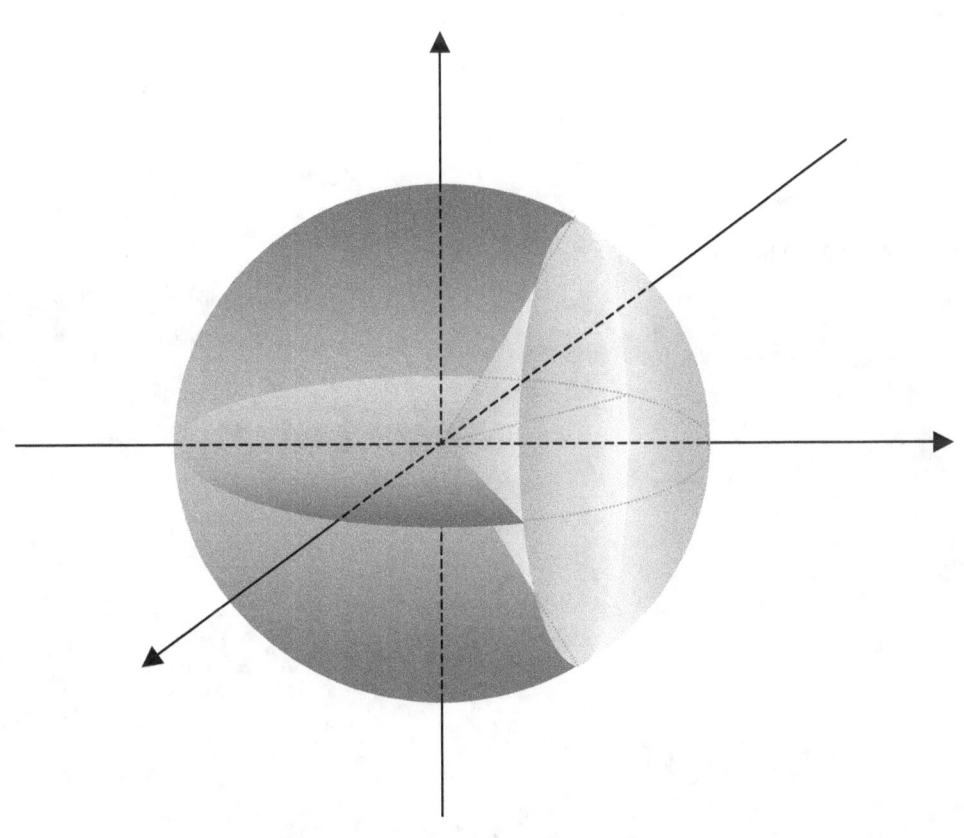

Seong R. KIM

Dear students:

Students need the best teacher, so you need examples, because examples are the best teacher. All the examples here are fully worked, and explain **how** the basic and essential tools in math are made, together with **what** they are, **how** they work, and **how** to work with them. Such tools include numbers, formulas, identities, equations, laws, etc.

Examples here begin with easy ones, of course. Covering every meter and yard properly, we can cover thousands of miles and kilometers. And it is particularly the case in math.

Of those examples therefore, some might even look too easy for you. It's not that easy though, to come up with those examples. Anyways, the bigger and the taller the tree, the deeper and the stronger the root.

Doing math, we work with ideas and run ideas, because every thing in math is an idea. A number is an idea, for instance, and the same is true for a line or circle, too. And putting ideas together, we build another, which becomes the base or an element of another, and each is connected. And that's the way your math grows. So you get to build a circuit, and sometimes, need to fill the gap or repair the circuit so that you get the sense of it.

So your calculation runs properly, and you get the problem solved.

The examples have been made and arranged so that they get tougher (or sometimes easier for some reason) as you proceed with them. In particular, similar examples with some variations are strategically repeated so that you can get the ideas or the tools tricky or complicated, and can get them mastered.

This book is however, nothing but a bunch of examples until you get it powered. How then, to get it powered, and make it run and work for you?

Just read it, and then, do each example in writing. And it is important to note that you do it in **your** writing. Just watching someone doing it, you just only feel that you can do it. If you do it, you can do it, but if you don't, we can hardly. It's a cliché, of course, but is always true that knowing is one thing and doing is another.

I've been helping students grow, take care of, and run their own math. The area covers algebra and geometry for high school or college students, and is especially for equations (for unknowns or curves), functions, and their graphs, which are the basic elements in calculus, which's been the core of my interest from my early age in high school.

Of my students, some are quite poor in math, and thus, are afraid of or hate math, some require special education because of exceptional intelligence, some are smart enough, some are naïve and diligent, some are clever but lazy, and most behave in general. All the students are badly after though, one thing in common: a strong and secure math skill. It is of course, the prime objective of my work, and I'm always happy to and eager to help them achieve it. The problem was however, that many of them wanted it to be purchased. And the question is, can we buy it?

We can buy the means, of course. And a solid math skill is feasible, too. We know however, we can't buy love, and the same is true for the math skill, too. It's not what we can buy or sell, and not what we can give or take. It is however, what we can grow, and need to grow. Your math grows as much as you grow and take care of it. So does mine.

What math then, do students most often do or use in high schools or colleges?

It is algebra and geometry. What algebra though?

Elementary algebra, of course
Doing the algebra, we work with numbers (many in kinds), constants, variables, ratios, rates, expressions, equations, inequalities, functions, identities, formulas, laws, etc., together with signs and symbols. And if we want to do algebra properly, we want to know their natures and how they mingle with each other.

So studying math ideas or tools, you want to know **what** they are, **how** they work, and **how** to work with them or **what** to do with them.   What then, about the geometry?

Basically, the geometry has much to do with shapes, positions, and angles. The shapes begin with triangles and circles, and move on to rectangles, squares, parallelograms or rhombuses, trapezoids, tetragons, other polygons, polyhedrons, etc.

Doing the geometry, too, though, we need to do the algebra stated above. So it is analytic geometry, often called coordinate geometry, too. And doing it, we can specify positions using coordinates. So in the geometry, basically, we work with graphs. Putting a math idea in a graph, we can not only effectively think about it but actually see it, too, and therefore, can efficiently work with it. What idea then, is it?

The idea begins with a point, line, parabola, circle, ellipse, and hyperbola, called a conic section or basic curve, and then, moves on to other curves, planes, surfaces, volumes, and other objects in various dimensional spaces, together with vectors.

And using an angle, we can specify an amount of turn or change in direction.

So learning, using, or applying those ideas or math tools, we get to solve problems.

And this book can help. It can help learn them, and use them so that you can navigate to find solutions to problems. And in particular, it can help come up with answers to those **what**s and **how**s stated above. So it can help you grow and run your own math, and thus, can help achieve your solid math skill.

It is however, not a magic book giving you a math skill of high caliber overnight. And it can have many mistakes, too. There is no magic, and math is full of facts and ideas. And it is after all, not me and not your teacher but you who put together some of those facts and ideas, and understand it. Putting facts and ideas together, understanding it, and taking care of what you have learned, you grow your math. And this book can help.

This is a book of examples designed to help you grow your math, and assumes that you are a real beginner. This book requires though, time and effort, the amount of which need to be substantial, too, but will be worth it. That's because you want a substantial achievement, and will get it. And probably, you will get to see this book helping you get there much faster than expected. And then, you will get to see the way math runs.

In math, everything is an idea. So is a problem. And solving it, we put it many different ways. For instance, while expanding or reducing it, or modifying or converting it, we keep searching for the solution, approaching the solution, and eventually, can get there. So don't look for the solution outside the problem. The solution is inside the problem if the problem is properly made.

If it is not, no solution is the solution. And in fact, it is often the case a problem itself is the solution. We can put a problem in many different ways, and eventually, can end up with the solution.   How come then, is the solution no other than the problem?

For instance, the solution to $3232 \div 101$ is 32.   And we can put it this way:

$$3232 \div 101 = \frac{3232}{101} = \frac{32 \times 101}{101} = \frac{32}{1} = 32 \implies 3232 \div 101 = 32.$$

And we can get this, too: $32 \implies 3232 \div 101$.   How?

$$32 = \frac{32}{1} = \frac{32 \times 101}{101} = \frac{3232}{101} = 3232/101 = 3232 \div 101.$$ Too easy?

For another instance, the solution to $ax^2 + bx + c = 0$ is: $x = \frac{-b \pm \sqrt{b^2 - 4ac}}{2a}$, which is called the quadratic formula.   How come then, is the solution no other than the problem?

We can put it this way:

$$x = \frac{-b \pm \sqrt{b^2-4ac}}{2a} \implies 2ax = -b \pm \sqrt{b^2 - 4ac} \implies 2ax + b = \pm\sqrt{b^2 - 4ac}$$

$$\implies (2ax + b)^2 = b^2 - 4ac \implies 4a^2x^2 + 4abx + b^2 = b^2 - 4ac$$

$$\implies 4a^2x^2 + 4abx = -4ac \implies ax^2 + bx = -c \implies ax^2 + bx + c = 0.$$

And we can get this, too: $ax^2 + bx + c = 0 \implies x = \frac{-b \pm \sqrt{b^2-4ac}}{2a}$.   How?

$$ax^2 + bx + c = a(x^2 + \tfrac{b}{a}x) + c = a(x^2 + \tfrac{b}{a}x + \tfrac{b^2}{4a^2} - \tfrac{b^2}{4a^2}) + c = a(x^2 + \tfrac{b}{a}x + \tfrac{b^2}{4a^2}) - \tfrac{b^2}{4a} + c$$

$$= a(x + \tfrac{b}{2a})^2 - \tfrac{b^2-4ac}{4a} = 0 \implies a(x + \tfrac{b}{2a})^2 = \tfrac{b^2-4ac}{4a} \implies (x + \tfrac{b}{2a})^2 = \tfrac{b^2-4ac}{4a^2} \implies x + \tfrac{b}{2a} = \pm\sqrt{\tfrac{b^2-4ac}{4a^2}}$$

$$\implies x = -\tfrac{b}{2a} \pm \tfrac{\sqrt{b^2-4ac}}{2a} = \tfrac{-b \pm \sqrt{b^2-4ac}}{2a} \implies x = \tfrac{-b \pm \sqrt{b^2-4ac}}{2a}.$$

And we call the set of processes above, algebra.

So if a problem is well defined, that is, if it makes sense, we should be able to get it solved the way below:

A problem ⇒ ... ⇒ ... ⇒ the solution, and thus: **the problem ⇒ the solution**.

So solving a problem, we put it many different ways so that we can get to the solution.

And that's the way, math runs.

May your math run very well.

Seong R. Kim

B.S. Math. Michigan Tech. Univ.   M.S. Math. Rensselaer Polytechnic Institute

Notes:

This book is one of three books about some basics in numbers, and covers arithmetic operations. And the operations are on numbers called real numbers. We can classify real numbers in two ways. In one, a number is an integer or a non-integer, which is not an integer. And in the other, a number is rational or irrational, which is not rational.

So if a number is not irrational, it is rational. Among rational numbers, we have integers and non-integers. And if a number is not an integer, it is a non-integer. Among non-integers, we have rational numbers and irrational numbers. It's confusing, isn't it?

So usually, we classify real numbers into three groups: integers, rational numbers, and irrational numbers. And this book is about rational and irrational numbers.

So the book covers non-integers, and helps understand what those are about and how they mingle with arithmetic operations so that you can not only use them properly but can develop your own idea to make use of them, too, solving problems, of course.

It doesn't just cover though, how to do arithmetic with rational or irrational numbers, that is, it doesn't merely explain how to apply to those numbers additions, subtractions, multiplications, and divisions. But it helps also understand the nature of non-integers.

So this book covers the concepts of those numbers, and helps you understand them, and use them efficiently as well as properly. And it can help speed up your calculation, too.

What then, about integers?

They are covered in two books. One is **Calculation Examples Arithmetic 1**. And the book covers integers, and helps understand what integers are about and how they mingle with arithmetic operations so that you can not only use integers properly but can develop your own idea to make use of them, too, solving problems, of course.

It doesn't just cover though, how to do arithmetic with integers, that is, how to do additions, subtractions, multiplications, and divisions. But it helps also understand the nature of integers and the nature of arithmetic operations as additions or divisions. And in particular, it covers the nature and the use of numbers called inverses. What inverse?

It can be called the negative or the reciprocal depending on the arithmetic operation. So this book covers those math ideas (concepts), and helps you understand them, and use them efficiently as well as properly. It can help speed up your calculation, too.

And the other is **Calculation Examples Arithmetic 2**, which is designed for those students who want to study word problems, where students need to manage integers and set up expressions. And you will get to see some nature of integers and some tools useful or handy using or working with integers. The tools are math ideas, called theorems or formulas, which you can use doing problems with multiples or divisors, etc.

And thus, And thus, all the three books will help you grab math ideas often used in real life as well as in math courses. The ideas are about real numbers, their arithmetic operations, and their nature, so you will get to see what those numbers are about and how they mingle with arithmetic operations, and will get to develop your own idea to make use of those, solving problems, of course.

In short, the books help you strengthen fundamentals in math to increase your skill of algebra, that is, calculation techniques, providing you with examples, showing all the steps and the ideas behind, and explaining what the math ideas are about.

Contents

1. Rational Numbers 1

To begin with, what are rational numbers?

The word 'rational' in 'rational number' is the adjective of <u>ratio</u>, and has little to do with the dictionary meaning of reasonable. What ratio then, is a rational number?

It can be called an *integer ratio*. More specifically, a rational number is <u>a ratio of an integer to another</u> as $\frac{1}{3}$, which can be taken as a ratio of 1 to 3.

So though it sounds obvious, 1 is a third of three, and 1 is three times a third, that is, $1 = 3 \times \frac{1}{3}$. And usually, we take $\frac{1}{3}$ as a third of 1 or one third or just a third.

Taking more instances of rational numbers, we have: $\frac{2}{5}$, $-\frac{3}{4}$, $0.2 = \frac{2}{10} = \frac{1}{5}$, etc.

And we can take 2/5 as a ratio of 2 to 5, and 2 is two fifths of 5, and 2 is five times 2/5, that is, $2 = 5 \times (2/5)$. Usually though, we take 2/5 as a number called two fifths.

So the nature of a rational number is an integer ratio, a ratio of an integer to another.

And rational numbers can be quickly just called <u>rationals</u>, and <u>include integers</u>, too.
So calling a number a rational, we mean it's a rational number.
Why is an integer a rational, too, though?

We can put every integer in an integer ratio, too, as $7 = \frac{7}{1}$, $-5 = \frac{-5}{1}$, $0 = \frac{0}{1}$, etc.
And thus, the set of all integers is a part of the set of all rational numbers.

So if expressing a number in terms of an integer ratio, we get a rational number.

Usually though, saying a rational number, we mean a fractional number as $\frac{2}{3}$, 1.3, etc.

So if expressing an amount that has a part of 1 as a half or a third, we use a rational number. For instance, $1.3 = 1 + 0.3$, which means 1 and three tenths.

What do we mean by though, three tenths?

Dividing 1 by 10, we don't just divide 1 into 10 parts.
Dividing 1 by 10, we divide 1 into 10 <u>equal</u> parts, and then, take one of those ten parts, and we call the one part one tenth.

So dividing 1 by 10, we get one tenth. And we can put it this way: $1 \div 10 = \frac{1}{10} = 1/10$.

What then, about dividing 3 by 10?

Dividing 3 by 10, we divide 3 into ten equal parts, and then, take one of the ten parts, and we call the one part three tenths, which is: 3/10.   How come?

We know 3 is three of 1s.

So we can do the same division 3 times, and the same division is the division of 1 by 10, that is, $1 \div 10$. And from each of the same divisions, we get 1 tenth, which is 1/10.

So dividing 3 by 10, we can do $(1 \div 10)$ three times, and then, put together the three of $(1/10)$s, that is, we get: $\frac{1}{10} + \frac{1}{10} + \frac{1}{10}$.

So we get three tenths, and can put it this way: 3/10, which is $\frac{3}{10}$

So 3/10 is three of $\frac{1}{10}$s. That is to say that we get: $\frac{3}{10} = 3 \times \frac{1}{10}$.

And by the same token, dividing 7 by 10, we can do 7 times the division of 1 by 10, that is, we can do $(1 \div 10)$ seven times, and then, add together the seven of $(1/10)$s.

Then, we get seven tenths, and write it this way: $\dfrac{7}{10}$ or 7/10.

What then, about using such a number as a ratio?

Assuming A is a third of B, we can put it this way: $A/B = 1/3$, since 1 is a third of 3.

And in that case, we can say that B is three times A.

So we get: $A/B = 1/3 \Rightarrow B/A = 3/1 = 3 \Rightarrow B = 3 \times A$, which means B is 3 times A.

And since A is a third of B, we can put it this way, too: $A/B = 2/6$, since 2 is a third of 6.

So we get: $A/B = 2/6 = 1/3$.

And again, since A is a third of B, we can put it this way, too: $A/B = 3/9$, since 3 is a third of 9.   So we get: $A/B = 3/9 = 2/6 = 1/3$.

And again, since A is a third of B, we can put it this way, too: $A/B = 4/12$, since 4 is a third of 12. So we get: $A/B = 1/3 = 2/6 = 3/9 = 4/12 = 5/15 \ldots$ And in fact, we have:

$$\frac{2}{6} = \frac{1 \times 2}{3 \times 2} = \frac{1}{3}, \ \frac{3}{9} = \frac{1 \times 3}{3 \times 3} = \frac{1}{3}, \ \frac{4}{12} = \frac{1 \times 4}{3 \times 4} = \frac{1}{3}, \ \frac{5}{15} = \frac{1 \times 5}{3 \times 5} = \frac{1}{3}, \text{ etc.}$$

Putting it in a slightly different form, we have:

$2/6 = (1 \times 2)/(3 \times 2) = 1/3, \ 3/9 = (1 \times 3)/(3 \times 3) = 1/3, \ 4/12 = (1 \times 4)/(3 \times 4) = 1/3,$ etc.

So we can see that multiplying by the same number both the numerator and denominator, we get the same ratio. (The same number cannot be 0, of course.)

<u>So the same ratios can look different.</u>

And we usually simplify a ratio so that the numerator and denominator do not share the same divisor other than 1.   What same divisor though?

In the case of 2/6, 2 is a divisor of 2, and also, is a divisor of 6, too.
In the case of 3/9, 3 is a divisor of 2, and also, is a divisor of 9, too.
In the case of 4/12, 4 is a divisor of 4, and also, is a divisor of 12, too.

Simplifying a ratio, we divide by the same integer both the numerator and denominator so that we get smaller integers for both the numerator and denominator keeping the value of the ratio intact. (And of course, the same integer cannot be 0.)

For instance, 4/6, 6/9, and 8/12 are the same ratios, and all get simplified to 2/3, which is thus, the simplest.

And when a ratio is in its simplest form as 2/3, we say that the numerator and denominator are <u>prime to each other</u>, and do not share any divisor other than 1.

Now, let's move on to arithmetic on rationals.

By the way, doing a multiplication, we don't usually use the operator x, and instead, we just use a dot · or nothing if no ambiguity is expected. So for instance, meaning 3 x 4, we often put it this way: 3·4, and we usually put A x B this way: AB.

So to begin with, we have:

$$\bullet \quad \frac{A}{B} = \frac{AC}{BC} = \frac{AE}{BE} = \frac{AXY}{BXY} = \dots, \text{ where } B, C, D, E, X, \text{ and } Y \neq 0.$$

And putting it in a slightly different form, we have:

• $A/B = AC/BC = AD/BD = AE/BE = AXY/BXY = \dots$ where $B, C, D, E, X,$ and $Y \neq 0$.

So other than 0, multiplying by the same both the numerator and the denominator, we get the same ratio, that is, the value of the fraction (or the ratio) does not change.

And because of the fact above, adding fractions, we do the addition the way below:

$$\bullet \quad \frac{A}{B} + \frac{C}{D} = \frac{AD}{BD} + \frac{BC}{BD} = \frac{AD + BC}{BD}, \text{ where } B \text{ and } D \neq 0.$$

And putting it a bit differently, we have:

$A/B + C/D = AD/BD + BC/BD = (AD + BC)/BD$, where B and $D \neq 0$, of course.

For instance: $\dfrac{2}{3} + \dfrac{5}{7} = \dfrac{2 \times 7}{3 \times 7} + \dfrac{5 \times 3}{7 \times 3} = \dfrac{14}{21} + \dfrac{15}{21} = \dfrac{14 + 15}{21} = \dfrac{29}{21}$.

And putting it in a slightly different form, we have:

2/3 + 5/7 = (2 x 7)/(3 x 7) + (5 x 3)/(7 x 3) = 14/21 + 15/21 = (14 + 15)/21 = 29/21.

And next, moving on to subtractions, we can do a subtraction the way below:

$$\bullet \quad \frac{A}{B} - \frac{C}{D} = \frac{AD}{BD} - \frac{BC}{BD} = \frac{AD - BC}{BD}, \text{where } \mathbf{B} \text{ and } \mathbf{D} \neq \mathbf{0}.$$

And putting it a bit differently, we have:

A/B – C/D = AD/BD – BC/BD = (AD – BC)/BD, where $\mathbf{B}$ and $\mathbf{D} \neq \mathbf{0}$, of course.

For instance: $\dfrac{2}{3} - \dfrac{5}{7} = \dfrac{2 \text{ x } 7}{3 \text{ x } 7} - \dfrac{5 \text{ x } 3}{7 \text{ x } 3} = \dfrac{14}{21} - \dfrac{15}{21} = \dfrac{14 - 15}{21} = \dfrac{-1}{21} = -\dfrac{1}{21}.$

And putting it a bit differently, we have:

2/3 – 5/7 = (2 x 7)/(3 x 7) – (5 x 3)/(7 x 3) = 14/21 – 15/21 = (14 – 15)/21 = -1/21.

So putting threads together, we have:

$$\frac{A}{B} + \frac{C}{D} = \frac{AD}{BD} + \frac{BC}{BD} = \frac{AD + BC}{BD}, \text{where } \mathbf{B} \text{ and } \mathbf{D} \neq \mathbf{0}.$$

$$\frac{A}{B} - \frac{C}{D} = \frac{AD}{BD} - \frac{BC}{BD} = \frac{AD - BC}{BD}, \text{where } \mathbf{B} \text{ and } \mathbf{D} \neq \mathbf{0}.$$

And putting the two above into one expression, we get:

$$\frac{A}{B} \pm \frac{C}{D} = \frac{AD}{BD} \pm \frac{BC}{BD} = \frac{AD \pm BC}{BD}, \text{where } \mathbf{B} \text{ and } \mathbf{D} \neq \mathbf{0}.$$

Examples 1 in Rationals

Find the larger of the two in each of the pairs below:

Note: show your work.

0. 1/2 and 3/4

1. 1/2 and 1/3

2. 3/5 and 3/7

3. 5/4 and 7/4

4. 3/4 and 7/9

5. 4/5 and 5/6

6. 3/5 and 5/7

7. 9/10 and 11/12

8. 7/5 and 5/3

9. 15/11 and 11/7

A. -3/5 and -2/5

B. -7/9 and 1/100

C. -11/12 and -15/16

D. $\frac{78943828}{78943880}$ and $\frac{987656789}{987656792}$

E. $\frac{78943828}{987656789}$ and $\frac{78943830}{987656792}$

Suggestions or Solutions
To the Examples 1 in Rationals

How can we see if one of two numbers is greater than the other?

Basically, we have two ways. One is taking the difference between the two, and the other is taking the ratio between the two.

First, if we subtract one from the other and the difference is positive, the one is greater.

So for instance, if we get: $A - B > 0$, we get: $A > B$. If $A - B = 0$, we get: $A = B$. And of course, if $A - B < 0$, we get: $A < B$.

Next, if the ratio of one to the other is greater than 1, the one is grater.

So for instance, if we get: $\frac{A}{B} > 1$, we get: $A > B$. If $\frac{A}{B} = 1$, we get: $A = B$. And of course, if $\frac{A}{B} < 1$, we get: $A < B$.

What then, about fractional numbers as 1/2 and 2/3?

If the denominators are the same, we can see which is bigger taking the difference or the ratio between the numerators. What if however, the denominators are different?

We can make the denominators the same. That is, taking the common denominator, and then, comparing the numerators using the ways above, we can see which is the bigger.

0.   1/2 and 3/4

1/2 = 2/4, so 3/4 is larger.

1.   1/2 and 1/3

We can see that the numerators are the same. The larger the denominator, the smaller the fraction if the fraction is positive. So 1/2 is the larger.

2.   3/5 and 3/7

We can see that the numerators are the same. The larger the denominator, the smaller the fraction if the fraction is positive. So 3/5 is the larger.

3.   5/4 and 7/4

The denominators are the same. The larger the numerator, the larger the fraction if the fraction is positive. So 7/4 is the larger.

4.   3/4 and 7/9

Taking the common denominator, we get: 3/4 = 27/36, and 7/9 = 28/36.
So 7/9 is the larger.

5.   4/5 and 5/6

4/5 = 24/30, and 5/6 = 25/30. So 5/6 is the larger.

6.   3/5 and 5/7

3/5 = 21/35, and 5/7 = 25/35. So 5/7 is the larger.

7.   9/10 and 11/12

9/10 = 108/120, and 11/12 = 110/120. So 11/12 is the larger.

And we can put it the way below, too:
1 − 9/10 = 1/10, 1 − 11/12 = 1/12, and 1/12 is smaller than 1/10. So 11/12 is the larger.

8.   7/5 and 5/3

7/5 = 21/15, and 5/3 = 25/15. So 5/3 is larger.

9.   15/11 and 11/7

15/11 = 105/77, and 11/7 = 121/77. So 11/7 is larger.

A.   -3/5 and -2/5

The larger the magnitude, the smaller the negative number. So -2/5 is larger.

B.   -7/9 and 1/100

A positive number is larger than negative number. So 1/100 is larger.

C.   -11/12 and -15/16

The difference between the numerator and the denominator is 1.
However, the numbers are negative. So -11/12 is the larger.

D. $\frac{78943828}{78943880}$ and $\frac{987656789}{987656792}$

For $\frac{78943828}{78943880}$, the difference between the denominator and the numerator is 2.

And for $\frac{987656789}{987656792}$, the difference is 3.

And we can see that both fractions are less than 1.

Next, the denominator in the second fraction is much bigger than the one in the first.

So we can say that the second fraction is much closer to 1 than the first.

And thus, $\frac{987656789}{987656792}$ is the larger. In fact:

$\frac{78943828}{78943880} = 0.99999997446655311757739648557715$

$\frac{987656789}{987656792} = 0.999999996962507599502236805353740$

And we can get the same the way below, too:

First, we have: $78943828 = 78943830 - 2$.

So we get: $\frac{78943828}{78943880} = \frac{78943830}{78943880} - \frac{2}{78943880} = 1 - \frac{2}{78943880}$.

Next. we have: $987656789 = 987656792 - 3$.

So we get: $\frac{987656789}{987656792} = 1 - \frac{3}{987656792}$.

And we can easily see that $\frac{2}{78943880} > \frac{3}{987656792}$, since the denominator in $\frac{2}{78943880}$ is much smaller, and both numerators are about the same. So we get:

$1 - \frac{2}{78943880} < 1 - \frac{3}{987656792} \implies \frac{78943828}{78943880} < \frac{987656789}{987656792}$.

E. $\frac{78943828}{987656789}$ and $\frac{78943830}{987656792}$

Looking at the denominators, we can see both are about the same.

And looking at the numerators, we can see both are about the same, too.

And taking the difference between the two, we get:

$$\frac{78943828}{987656789} - \frac{78943830}{987656792} = \frac{78943828 \times 987656792}{987656789 \times 987656792} - \frac{78943830 \times 987656789}{987656789 \times 987656792}$$

$$= \frac{78943828 \times 987656792 - 78943830 \times 987656789}{987656789 \times 987656792}$$

So without using a calculator, we cannot see quickly enough if is positive, 0, or negative. What then, can we do?

To begin with, we can have: $78943828 = 78943830 - 2$

So we get:

$78943828 \times 987656792 - 78943830 \times 987656789$

$= (78943830 - 2)987656792 - 78943830 \times 987656789$

$= 78943830 \times 987656792 - 2 \times 987656792 - 78943830 \times 987656789$

$= 78943830(987656792 - 987656789) - 2 \times 987656792$

$= 78943830 \times 3 - 2 \times 987656792$

And next, we can see that:

$78943830 \times 3 < 80000000 \times 3$, which has 9 digits, and 2×987656792 has 10 digits.

So we can see that $78943830 \times 3 - 2 \times 987656792 < 0$.

That is, the difference is negative. And thus, $\frac{78943830}{987656792}$ is the larger.

Examples 2 in Rationals

0. $\dfrac{3}{7}+\dfrac{4}{9}$

1. $\dfrac{3}{7}-\dfrac{4}{9}$

2. $\dfrac{12}{15}+\dfrac{19}{23}$

3. $\dfrac{12}{15}-\dfrac{19}{23}$

4. $\dfrac{8}{9}+\dfrac{11}{12}$

5. $\dfrac{11}{12}-\dfrac{8}{9}$

6. $\dfrac{12}{35}+\dfrac{4}{15}$

7. $\dfrac{12}{27}-\dfrac{132}{15}$

8. $\dfrac{35}{225}+\dfrac{105}{625}$

9. $\dfrac{45}{1225}-\dfrac{70}{75}$

A. $\dfrac{13}{19}+\dfrac{23}{247}$

B. $\dfrac{23}{47}-\dfrac{47}{23}$

C. $\dfrac{24}{88}+\dfrac{32}{36}$

D. $\dfrac{72}{216}-\dfrac{42}{252}$

E. $\dfrac{60}{105}+\dfrac{252}{168}$

F. $\dfrac{144}{468}-\dfrac{54}{244}$

G. $\dfrac{3}{4}+\dfrac{4}{3}+2$

H. $\dfrac{4}{5}+\dfrac{5}{4}-2$

I. $\dfrac{13}{29}+\dfrac{29}{13}-2$

J. $\dfrac{161}{101}-\dfrac{101}{161}$

K. $\dfrac{51}{61}+\dfrac{61}{51}+2$

L. $\dfrac{151}{161}-\dfrac{1510}{1610}$

M. $\dfrac{11}{111}+\dfrac{111}{11}+\dfrac{3}{4}+\dfrac{3}{8}+\dfrac{3}{16}+\dfrac{22}{32}$

N. $\dfrac{22}{111}-\dfrac{111}{22}$

O. $\dfrac{1010}{101}+\dfrac{1}{10}$

P. $\dfrac{2525}{101}-\dfrac{3}{25}$

Q. $\dfrac{675675}{1001}+\dfrac{4040}{101}$

R. $\dfrac{12321}{111}-\dfrac{1210}{11}$

Suggestions or Solutions
To the Examples 2 in Rationals

0. $\dfrac{3}{7}+\dfrac{4}{9}=\dfrac{3\cdot 9}{7\cdot 9}+\dfrac{4\cdot 7}{9\cdot 7}=\dfrac{27}{63}+\dfrac{28}{63}=\dfrac{27+28}{63}=\dfrac{55}{63}$

1. $\dfrac{3}{7}-\dfrac{4}{9}=\dfrac{3\cdot 9}{7\cdot 9}-\dfrac{4\cdot 7}{9\cdot 7}=\dfrac{27}{63}-\dfrac{28}{63}=\dfrac{27-28}{63}=\dfrac{-1}{63}=-\dfrac{1}{63}$

2. $\dfrac{12}{15}+\dfrac{19}{23}=\dfrac{4\cdot 3}{5\cdot 3}+\dfrac{19}{23}=\dfrac{4}{5}+\dfrac{19}{23}=\dfrac{4\cdot 23}{5\cdot 23}+\dfrac{19\cdot 5}{23\cdot 5}=\dfrac{92}{115}+\dfrac{95}{115}=\dfrac{92+95}{115}=\dfrac{187}{115}$

3. $\dfrac{12}{15}-\dfrac{19}{23}=\dfrac{4\cdot 3}{5\cdot 3}-\dfrac{19}{23}=\dfrac{4}{5}-\dfrac{19}{23}=\dfrac{4\cdot 23}{5\cdot 23}-\dfrac{19\cdot 5}{23\cdot 5}=\dfrac{92}{115}-\dfrac{95}{115}=\dfrac{92-95}{115}=\dfrac{-3}{115}=-\dfrac{3}{115}$

4. $\dfrac{8}{9}+\dfrac{11}{12}=\dfrac{8\cdot 12}{9\cdot 12}+\dfrac{11\cdot 9}{12\cdot 9}=\dfrac{96}{108}+\dfrac{99}{108}=\dfrac{96+99}{108}=\dfrac{195}{108}=\dfrac{65\cdot 3}{36\cdot 3}=\dfrac{65}{36}$

And we can put it the way below, too:

First, we can get: $\dfrac{8}{9}+\dfrac{1}{9}=1\Rightarrow\dfrac{8}{9}=1-\dfrac{1}{9}$, and $\dfrac{11}{12}+\dfrac{1}{12}=1\Rightarrow\dfrac{11}{12}=1-\dfrac{1}{12}$.

You could do it by heart.

Then, we can get: $\dfrac{8}{9}+\dfrac{11}{12}=1-\dfrac{1}{9}+1-\dfrac{1}{12}=2-(\dfrac{1}{9}+\dfrac{1}{12})$.

Meanwhile, $\dfrac{1}{9}+\dfrac{1}{12}=\dfrac{4}{9\cdot 4}+\dfrac{3}{12\cdot 3}=\dfrac{4}{36}+\dfrac{3}{36}=\dfrac{7}{36}$.

So we get: $\dfrac{8}{9}+\dfrac{11}{12}=2-(\dfrac{1}{9}+\dfrac{1}{12})=2-\dfrac{7}{36}=\dfrac{72-7}{36}=\dfrac{65}{36}$

5. $\dfrac{11}{12} - \dfrac{8}{9} = \dfrac{11\cdot 3}{12\cdot 3} - \dfrac{8\cdot 4}{9\cdot 4} = \dfrac{33}{36} - \dfrac{32}{36} = \dfrac{1}{36}.$

And we can do it the way below, too, of course.

$\dfrac{11}{12} - \dfrac{8}{9} = \dfrac{11\cdot 9}{12\cdot 9} - \dfrac{8\cdot 12}{9\cdot 12} = \dfrac{99}{108} - \dfrac{96}{108} = \dfrac{3}{108} = \dfrac{3}{36\cdot 3} = \dfrac{1}{36}.$

6. $\dfrac{12}{35} + \dfrac{4}{15}$

First, we can get: $\dfrac{12}{35} + \dfrac{4}{15} = \dfrac{4\cdot 3}{35} + \dfrac{4\cdot 1}{15} = 4(\dfrac{3}{35} + \dfrac{1}{15}).$

And next, we get: $\dfrac{3}{35} + \dfrac{1}{15} = \dfrac{3\cdot 3}{35\cdot 3} + \dfrac{1\cdot 7}{15\cdot 7} = \dfrac{9}{105} + \dfrac{7}{105} = \dfrac{16}{105}.$

So we get: $\dfrac{12}{35} + \dfrac{4}{15} = 4(\dfrac{3}{35} + \dfrac{1}{15}) = 4\cdot\dfrac{16}{105} = \dfrac{64}{105}.$

Of course, we can get the same result the way below, too:

$\dfrac{12}{35} + \dfrac{4}{15} = \dfrac{12\cdot 15}{35\cdot 15} + \dfrac{4\cdot 35}{15\cdot 35} = \ldots$

7. $\dfrac{12}{27} - \dfrac{132}{15}$

We have: $\dfrac{12}{27} = \dfrac{3\cdot 4}{3\cdot 9} = \dfrac{4}{9}$, and $\dfrac{132}{15} = \dfrac{3\cdot 44}{3\cdot 5} = \dfrac{44}{5}.$

So we get: $\dfrac{12}{27} - \dfrac{132}{15} = \dfrac{4}{9} - \dfrac{44}{5} = 4(\dfrac{1}{9} - \dfrac{11}{5})$, and $\dfrac{1}{9} - \dfrac{11}{5} = \dfrac{5-99}{45} = \dfrac{-94}{45}.$

Thus, we get: $\dfrac{12}{27} - \dfrac{132}{15} = 4(\dfrac{1}{9} - \dfrac{11}{5}) = 4\cdot(-\dfrac{94}{45}) = -\dfrac{376}{45}$

And of course, we can do it the way below, too:

$\dfrac{12}{27} - \dfrac{132}{15} = \dfrac{12\cdot 15}{27\cdot 15} - \dfrac{132\cdot 27}{15\cdot 27} = \ldots$

8. $\dfrac{35}{225}+\dfrac{105}{625}$

To begin with, we get: $\dfrac{35}{225}+\dfrac{105}{625}=\dfrac{7\cdot5}{45\cdot5}+\dfrac{21\cdot5}{125\cdot5}=\dfrac{7}{45}+\dfrac{21}{125}=7(\dfrac{1}{45}+\dfrac{3}{125})$.

And we can get: $\dfrac{1}{45}+\dfrac{3}{125}=\dfrac{1}{9\cdot5}+\dfrac{3}{25\cdot5}=\dfrac{1\cdot25}{9\cdot5\cdot25}+\dfrac{3\cdot9}{25\cdot5\cdot9}=\dfrac{25+27}{1125}=\dfrac{52}{1125}$.

So we get: $\dfrac{35}{225}+\dfrac{105}{625}=7(\dfrac{1}{45}+\dfrac{3}{125})=7\cdot\dfrac{52}{1125}=\dfrac{364}{1125}$.

9. $\dfrac{45}{1225}-\dfrac{70}{75}$

First, we have: $45 = 5 \times 9 = 5 \times 3^2$, and $1225 = 5 \times 245 = 5 \times 5 \times 49 = 5^2 \times 7^2$.

So we get: $\dfrac{45}{1225}=\dfrac{5\cdot3^2}{5^2 7^2}=\dfrac{3^2}{5\cdot7^2}$.

Next, we have: $70 = 7 \times 2 \times 5$, and $75 = 5 \times 15 = 5 \times 5 \times 3 = 5^2 \times 3$.

So we get: $\dfrac{70}{75}=\dfrac{2\cdot7}{3\cdot5}$.

Thus, we get: $\dfrac{45}{1225}-\dfrac{70}{75}=\dfrac{3^2}{5\cdot7^2}-\dfrac{2\cdot7}{3\cdot5}=\dfrac{3^2\cdot3}{5\cdot7^2\cdot3}-\dfrac{2\cdot7\cdot7^2}{3\cdot5\cdot7^2}=\dfrac{27-686}{3\cdot5\cdot7^2}=\dfrac{-659}{735}$.

A. $\dfrac{13}{19}+\dfrac{23}{247}=\dfrac{13}{19}+\dfrac{23}{13\cdot19}=\dfrac{13\cdot13}{19\cdot13}+\dfrac{23}{13\cdot19}=\dfrac{169+23}{19\cdot13}=\dfrac{192}{247}$.

B. $\dfrac{23}{47}-\dfrac{47}{23}=\dfrac{23^2-47^2}{47\cdot23}=\dfrac{529-2209}{1081}=-\dfrac{1680}{1081}$.

C. $\dfrac{24}{88}+\dfrac{32}{36}=\dfrac{3\cdot8}{11\cdot8}+\dfrac{4\cdot8}{4\cdot9}=\dfrac{3}{11}+\dfrac{8}{9}=\dfrac{27+88}{99}=\dfrac{115}{99}$.

D. $\dfrac{72}{216} - \dfrac{42}{252}$

First, we have: $72 = 8 \times 9 = 2^3 \times 3^2$, and $216 = 2 \times 108 = 2^2 \times 54 = 2^2 \times 6 \times 9 = 2^3 \times 3^3$.

So we get: $\dfrac{72}{216} = \dfrac{1}{3}$.

Next, we have: $42 = 2 \times 3 \times 7$, and $252 = 2 \times 126 = 2 \times 2 \times 63 = 2^2 \times 3^2 \times 7$

So we get: $\dfrac{42}{252} = \dfrac{1}{6}$.

Thus, we get: $\dfrac{72}{216} - \dfrac{42}{252} = \dfrac{1}{3} - \dfrac{1}{6} = \dfrac{2-1}{6} = \dfrac{1}{6}$.

E. $\dfrac{60}{105} + \dfrac{252}{168}$

We have: $60 = 5 \times 12 = 2^2 \times 3 \times 5$, and $105 = 5 \times 21 = 3 \times 5 \times 7$

So we get: $\dfrac{60}{105} = \dfrac{4}{7}$.

Next, we have: $252 = 2^2 \times 3^2 \times 7$, and $168 = 2 \times 84 = 2^2 \times 42 = 2^3 \times 21 = 2^3 \times 3 \times 7$

So we get: $\dfrac{252}{168} = \dfrac{3}{2}$.

Thus, we get: $\dfrac{60}{105} + \dfrac{252}{168} = \dfrac{4}{7} + \dfrac{3}{2} = \dfrac{4 \cdot 2 + 3 \cdot 7}{14} = \dfrac{29}{14}$.

F. $\dfrac{144}{468} - \dfrac{54}{244}$

We have: $144 = 2 \cdot 72 = 2^2 36 = 2^4 3^2 = 12^2 = (4 \cdot 3)^2 = (2^2 3)^2 = 2^4 3^2$, and
$468 = 2 \cdot 234 = 2^2 117 = 2^2 3 \cdot 39 = 2^2 3^2 13$.

So we get: $\dfrac{144}{468} = \dfrac{2^2}{13}$.

Next, we have: $54 = 2 \cdot 27 = 2 \cdot 3^3$, and $244 = 2 \cdot 122 = 2^2 61$.

So we get: $\dfrac{54}{244} = \dfrac{3^3}{2 \cdot 61}$.

Thus, we get: $\dfrac{144}{468} - \dfrac{54}{244} = \dfrac{2^2}{13} - \dfrac{3^3}{2 \cdot 61} = \dfrac{2^2 2 \cdot 61}{2 \cdot 13 \cdot 61} - \dfrac{3^3 13}{2 \cdot 3 \cdot 61} = \dfrac{2^3 \cdot 61 - 3^3 13}{2 \cdot 3 \cdot 61} = \dfrac{137}{1586}$.

G. $\dfrac{3}{4}+\dfrac{4}{3}+2$

First, we get: $\dfrac{3}{4}+\dfrac{4}{3}=\dfrac{3^2+4^2}{12}$, and $2=\dfrac{2\cdot3\cdot4}{12}$.

So we get: $\dfrac{3}{4}+\dfrac{4}{3}+2=\dfrac{3^2+4^2}{12}+\dfrac{2\cdot3\cdot4}{12}=\dfrac{25+24}{12}=\dfrac{49}{12}$.

Note that we can have: $3^2+2\cdot3\cdot4+4^2=(3+4)^2$.

H. $\dfrac{4}{5}+\dfrac{5}{4}-2=\dfrac{4^2+5^2}{20}-\dfrac{2\cdot4\cdot5}{20}=\dfrac{16+25-40}{20}=\dfrac{1}{20}$

Note that we can have: $4^2-2\cdot4\cdot5+5^2=(4-5)^2=(5-4)^2$.

I. $\dfrac{13}{29}+\dfrac{29}{13}-2=\dfrac{13^2+29^2}{29\cdot13}-\dfrac{2\cdot13\cdot29}{29\cdot13}=\dfrac{16^2}{377}=\dfrac{256}{377}$.

Note that we can have: $13^2+29^2-2\text{ x }13\text{ x }29=(29-13)^2=(13-29)^2=16^2$.

J. $\dfrac{161}{101}-\dfrac{101}{161}=\dfrac{161^2-101^2}{101\cdot161}=\dfrac{25921-10201}{16261}=\dfrac{15720}{16261}$.

K. $\dfrac{51}{61}+\dfrac{61}{51}+2=\dfrac{51^2+61^2+2\cdot51\cdot61}{61\cdot51}=\dfrac{(51+61)^2}{61\cdot51}=\dfrac{112^2}{3111}=\dfrac{12544}{3111}$.

L. $\dfrac{151}{161}-\dfrac{1510}{1610}$

First, we have: $1510 = 151$ x 10 and $1610 = 161$ x 10. So we get: $\dfrac{1510}{1610}=\dfrac{151}{161}$.

Thus, we get: $\dfrac{151}{161}-\dfrac{1510}{1610}=0$.

M. $\quad \dfrac{11}{111} + \dfrac{111}{11} + \dfrac{3}{4} + \dfrac{3}{8} + \dfrac{3}{16} + \dfrac{22}{32}$

First, we have: $\dfrac{3}{4} = \dfrac{24}{32}, \ \dfrac{3}{8} = \dfrac{12}{32}, \ \text{and} \ \dfrac{3}{16} = \dfrac{6}{32}.$

So we get: $\dfrac{3}{4} + \dfrac{3}{8} + \dfrac{3}{16} + \dfrac{22}{32} = \dfrac{24 + 12 + 6 + 22}{32} = \dfrac{64}{32} = 2.$

And thus, we get:

$$\dfrac{11}{111} + \dfrac{111}{11} + \dfrac{3}{4} + \dfrac{3}{8} + \dfrac{3}{16} + \dfrac{22}{32} = \dfrac{11}{111} + \dfrac{111}{11} + 2 = \dfrac{(11+111)^2}{111 \cdot 11} = \dfrac{122^2}{1221} = \dfrac{14884}{1221}.$$

Note that we can have: $11^2 + 111^2 + 2 \times 11 \times 111 = (11 + 111)^2 = 122^2.$

N. $\quad \dfrac{22}{111} - \dfrac{111}{22} = \dfrac{22^2 - 111^2}{111 \cdot 22} = \dfrac{484 - 12321}{2442} = -\dfrac{11837}{2442}.$

O. $\quad \dfrac{1010}{101} + \dfrac{1}{10} = \dfrac{101 \cdot 10}{101} + \dfrac{1}{10} = 10 + \dfrac{1}{10} = \dfrac{100 + 1}{10} = \dfrac{101}{10}.$

P. $\quad \dfrac{2525}{101} - \dfrac{3}{25} = \dfrac{25 \cdot 101}{101} - \dfrac{3}{25} = 25 - \dfrac{3}{25} = \dfrac{25^2 - 3}{25} = \dfrac{625 - 3}{25} = \dfrac{622}{25}.$

Q. $\quad \dfrac{675675}{1001} + \dfrac{4040}{101}$

We have: $675675 = 675 \times 1001$, and $4040 = 4 \times 1010 = 40 \times 101$

So we get: $\dfrac{675675}{1001} + \dfrac{4040}{101} = 675 + 40 = 715.$

R. $\quad \dfrac{12321}{111} - \dfrac{1210}{11}$

We have: $12321 = 111^2$, and $121 = 11^2$

So we get: $\dfrac{12321}{111} - \dfrac{1210}{11} = \dfrac{111^2}{111} - \dfrac{11^2 10}{11} = 111 - 11 \cdot 10 = 111 - 110 = 1.$

2. Rational Numbers 2

To begin with, going over briefly the ideas covered in the previous section, we have:

- $\dfrac{A}{B} = \dfrac{AC}{BC} = \dfrac{AE}{BE} = \dfrac{AXY}{BXY} = \dots$, where B, C, D, E, X, and $Y \neq 0$.

So other than 0, multiplying by the same both the numerator and the denominator, we get the same ratio, that is, the value of the fraction (or the ratio) does not change.

And because of the fact above, adding fractions, we can do additions the way below:

- $\dfrac{A}{B} + \dfrac{C}{D} = \dfrac{AD}{BD} + \dfrac{BC}{BD} = \dfrac{AD+BC}{BD}$, where B and $D \neq 0$.

For instance: $\dfrac{2}{3} + \dfrac{5}{7} = \dfrac{2 \times 7}{3 \times 7} + \dfrac{5 \times 3}{7 \times 3} = \dfrac{14}{21} + \dfrac{15}{21} = \dfrac{14+15}{21} = \dfrac{29}{21}$.

And next, moving on to subtractions, we can do a subtraction the way below:

- $\dfrac{A}{B} - \dfrac{C}{D} = \dfrac{AD}{BD} - \dfrac{BC}{BD} = \dfrac{AD-BC}{BD}$, where B and $D \neq 0$.

For instance: $\dfrac{2}{3} - \dfrac{5}{7} = \dfrac{2 \times 7}{3 \times 7} - \dfrac{5 \times 3}{7 \times 3} = \dfrac{14}{21} - \dfrac{15}{21} = \dfrac{14-15}{21} = \dfrac{-1}{21} = -\dfrac{1}{21}$.

So putting threads together, we have:

$$\frac{A}{B} + \frac{C}{D} = \frac{AD}{BD} + \frac{BC}{BD} = \frac{AD+BC}{BD}, \text{ where } B \text{ and } D \neq 0.$$

$$\frac{A}{B} - \frac{C}{D} = \frac{AD}{BD} - \frac{BC}{BD} = \frac{AD-BC}{BD}, \text{ where } B \text{ and } D \neq 0.$$

And putting the two above into one expression, we get:

$$\frac{A}{B} \pm \frac{C}{D} = \frac{AD}{BD} \pm \frac{BC}{BD} = \frac{AD \pm BC}{BD}, \text{ where } B \text{ and } D \neq 0.$$

Now, moving next, on to multiplications, and multiplying, for instance, 1 by 1/3, what do we get?

We get: 1 x 1/3 = 1/3.   And we get: 2 x 1/ 3 = 2/3, 3 x 1/3 = 3/3 = 1, 4 x 1/3 = 4/3, etc.

What do we mean by though, for instance, 1 x 1/3 = 1/3?

Taking 1 of (1/3)s, we can multiply 1 by 1/3, and get 1/3.

Taking 2 of (1/3)s, we can multiply 2 by 1/3, and get 2/3.

Taking 3 of (1/3)s, we can multiply 3 by 1/3, and get 3/3 = 1.

And taking 4 of (1/3)s, we can multiply 4 by 1/3, and get 4/3.

And we know: $1 \times \frac{1}{3} = \frac{1}{3} \times 1$, $2 \times \frac{1}{3} = \frac{1}{3} \times 2$, $3 \times \frac{1}{3} = \frac{1}{3} \times 3$, and $4 \times \frac{1}{3} = \frac{1}{3} \times 4$.

What then, do we mean by $1 \times \frac{1}{3} = \frac{1}{3}$, for instance?

Taking a third of 1, we can multiply 1/3 by 1, and get 1/3.

Taking a third of 2, we can multiply 1/3 by 2, and get 2/3.

Taking a third of 3, we can multiply 1/3 by 3, and get 3/3 = 1.

And taking a third of 4, we can multiply 1/3 by 4, and get 4/3.

What then, do we mean by 2/3 x 5/7, that is, $\frac{2}{3}$ x $\frac{5}{7}$, for instance?

We know 2/3 is 2 of (1/3)s.

And we know taking $\frac{1}{3}$ x $\frac{5}{7}$, we can say that we take a third of 5/7.

So taking $\frac{2}{3}$ x $\frac{5}{7}$, we take 2 x $\frac{1}{3}$ x $\frac{5}{7}$, so we can say that we take twice a third of 5/7.

What then, is a third of 5/7?

We know 5/7 = 15/21, and a third of 15/21 is 5/21, since 15/21 is 15 of (1/21)s, and a third of 15 is 5.

So a third of 5/7, that is, $\frac{1}{3}$ x $\frac{5}{7}$ is: 5/21.

And we know multiplying 2/3 by 5/7, that is, taking $\frac{2}{3}$ x $\frac{5}{7}$, we take twice a third of 5/7.

So we get: $\frac{2}{3}$ x $\frac{5}{7}$ = 2 x ($\frac{1}{3}$ x $\frac{5}{7}$) = 2 x $\frac{5}{21}$, which is 10/21.

What then, do we mean by $\frac{5}{7}$ x $\frac{2}{3}$?

We know 5/7 is 5 of (1/7)s.
And we know taking $\frac{1}{7}$ x $\frac{2}{3}$, we can say that we take a seventh of 2/3.

So taking $\frac{5}{7}$ x $\frac{2}{3}$, we take 5 x $\frac{1}{7}$ x $\frac{2}{3}$, so we can say that we take 5 times a seventh of 2/3.

And we know 2/3 = 14/21, and a seventh of 14/21 is 2/21, since 14/21 is 14 of (1/21)s, and a seventh of 14 is 2.
So a seventh of 2/3, that is, $\frac{1}{7}$ x $\frac{2}{3}$ is 2/21.

And we know taking $\frac{5}{7}$ x $\frac{2}{3}$, we take 5 times a seventh of 2/3.

So we get: 5 x $\frac{1}{7}$ x $\frac{2}{3}$ = 5 x $\frac{2}{21}$, which is 10/21, which is the same as $\frac{5}{7}$ x $\frac{2}{3}$ = $\frac{3}{2}$ x $\frac{5}{7}$.

Now, how then, do we get 10 and 21 in the fraction 10/21 above?

We get 10 multiplying 5 by 2, and get 21 multiplying 7 by 3. Specifically:

• Taking the product of the numerators of 5/7 and 2/3, we get 10, which is the numerator of 10/21, which is the result of $\frac{5}{7}$ x $\frac{2}{3}$.

• Taking the product of the denominators of 5/7 and 2/3, we get 21, which is the denominator of 10/21.

So we can notice that taking a product of fractions, we get a fraction where the numerator is the product of the numerators, and the denominator is the product of the denominators of the fractions multiplied.

In short, we get: *A/B* x *C/D* = *AC/BD*, that is, we get: $\dfrac{A}{B}$ x $\dfrac{C}{D}$ = $\dfrac{AC}{BD}$.

And we get: $\dfrac{A}{B}$ x $\dfrac{C}{D}$ x $\dfrac{E}{F}$ = $\dfrac{ACE}{BDF}$, because $\dfrac{AC}{BD}$ x $\dfrac{E}{F}$ = $\dfrac{ACE}{BDF}$. And so forth

In the next section, we will move on to the next operations, divisions with rational numbers, after doing some more examples on additions, shown in the next pages.

Examples 3 in Rationals

0. $\dfrac{1}{2}+\dfrac{2}{3}+\dfrac{3}{4}+\dfrac{4}{5}$

1. $\dfrac{1}{2}-\dfrac{3}{4}+\dfrac{5}{6}-\dfrac{7}{8}$

2. $\dfrac{8}{9}-\dfrac{7}{8}+\dfrac{17}{18}-\dfrac{15}{16}$

3. $\dfrac{2}{3}+\dfrac{8}{9}+\dfrac{26}{27}+\dfrac{80}{81}$

4. $\dfrac{1}{4}+\dfrac{1}{16}+\dfrac{1}{64}+\dfrac{1}{256}$

5. $\dfrac{1}{2}+\dfrac{1}{3}+\dfrac{1}{4}+\dfrac{1}{5}$

6. $\dfrac{1}{10}+\dfrac{1}{100}+\dfrac{1}{1000}+\dfrac{1}{10000}$

7. $\dfrac{1}{10}-\dfrac{1}{100}-\dfrac{1}{1000}-\dfrac{1}{10000}$

8. $\dfrac{12}{20}+\dfrac{20}{30}+\dfrac{30}{42}+\dfrac{42}{56}$

9. $\dfrac{121}{11}+\dfrac{144}{12}+\dfrac{169}{13}+\dfrac{196}{14}$

A. $\dfrac{1}{1}-\dfrac{1}{11}-\dfrac{11}{121}-\dfrac{1331}{121}-\dfrac{14641}{1331}$

B. $\dfrac{55}{111}+\dfrac{333}{12321}+\dfrac{444}{24642}$

C. $\dfrac{2}{3}+\dfrac{3}{2}+\dfrac{3}{4}+\dfrac{4}{3}+\dfrac{4}{5}+\dfrac{5}{4}$

D. $\dfrac{15}{12}+\dfrac{20}{15}+\dfrac{35}{28}+\dfrac{36}{27}$

Suggestions or Solutions
To the Examples 3 in Rationals

0. $\dfrac{1}{2}+\dfrac{2}{3}+\dfrac{3}{4}+\dfrac{4}{5}$

We can get first: $\dfrac{1}{2}+\dfrac{3}{4}=\dfrac{2}{4}+\dfrac{3}{4}=\dfrac{2+3}{4}=\dfrac{5}{4}$.

And next, we can get: $\dfrac{2}{3}+\dfrac{4}{5}=\dfrac{2\cdot5}{3\cdot5}+\dfrac{3\cdot4}{3\cdot5}=\dfrac{10}{15}+\dfrac{12}{15}=\dfrac{10+12}{15}=\dfrac{22}{15}$.

So we get: $\dfrac{1}{2}+\dfrac{2}{3}+\dfrac{3}{4}+\dfrac{4}{5}=\dfrac{5}{4}+\dfrac{22}{15}=\dfrac{5\cdot15}{60}+\dfrac{22\cdot4}{60}=\dfrac{75+88}{60}=\dfrac{163}{60}$

1. $\dfrac{1}{2}-\dfrac{3}{4}+\dfrac{5}{6}-\dfrac{7}{8}$

To begin with, we can get firt: $\dfrac{1}{2}-\dfrac{3}{4}-\dfrac{7}{8}=\dfrac{4}{8}-\dfrac{6}{8}-\dfrac{7}{8}=\dfrac{4-6-7}{8}=-\dfrac{9}{8}$.

So we get:

$$\dfrac{1}{2}-\dfrac{3}{4}+\dfrac{5}{6}-\dfrac{7}{8}=\dfrac{5}{6}-\dfrac{9}{8}$$

$$=\dfrac{5}{6}-1-\dfrac{1}{8}\qquad\{\ \dfrac{9}{8}=\dfrac{8}{8}+\dfrac{1}{8}=1+\dfrac{1}{8}\ \}$$

$$=\dfrac{5}{6}-\dfrac{5}{6}-\dfrac{1}{6}-\dfrac{1}{8}\qquad\{\ -1=-\dfrac{5}{6}-\dfrac{1}{6}\ \}$$

$$-\dfrac{1}{6}-\dfrac{1}{8}=-\dfrac{4}{24}-\dfrac{3}{24}=-\dfrac{7}{24}$$

2. $\dfrac{8}{9} - \dfrac{7}{8} + \dfrac{17}{18} - \dfrac{15}{16}$

We can get first: $\dfrac{8}{9} + \dfrac{17}{18} = \dfrac{16}{18} + \dfrac{17}{18} = \dfrac{33}{18},$ and $-\dfrac{7}{8} - \dfrac{15}{16} = -\dfrac{14}{16} - \dfrac{15}{16} = -\dfrac{29}{16}.$

So we get:

$$\dfrac{8}{9} - \dfrac{7}{8} + \dfrac{17}{18} - \dfrac{15}{16} = \dfrac{33}{18} - \dfrac{29}{16} = \dfrac{33 \cdot 8}{18 \cdot 8} - \dfrac{29 \cdot 9}{16 \cdot 9} = \dfrac{264}{144} - \dfrac{180 + 81}{144} = \dfrac{264 - 261}{144} = \dfrac{3}{144}$$

3. $\dfrac{2}{3} + \dfrac{8}{9} + \dfrac{26}{27} + \dfrac{80}{81}$

To begin with, $\dfrac{2}{3} = \dfrac{2 \cdot 27}{3 \cdot 27} = \dfrac{54}{81}, \quad \dfrac{8}{9} = \dfrac{8 \cdot 9}{9 \cdot 9} = \dfrac{72}{81},$ and $\dfrac{26}{27} = \dfrac{26 \cdot 3}{27 \cdot 3} = \dfrac{78}{81}.$

So we get:

$$\dfrac{2}{3} + \dfrac{8}{9} + \dfrac{26}{27} + \dfrac{80}{81} = \dfrac{54}{81} + \dfrac{72}{81} + \dfrac{78}{81} + \dfrac{80}{81} = \dfrac{126 + 158}{81} = \dfrac{284}{81}$$

4. $\dfrac{1}{4} + \dfrac{1}{16} + \dfrac{1}{64} + \dfrac{1}{256}$

First, we can get: $\dfrac{1}{4} = \dfrac{64}{4 \cdot 64} = \dfrac{64}{256}, \quad \dfrac{1}{16} = \dfrac{16}{16 \cdot 16} = \dfrac{16}{256},$ and $\dfrac{1}{64} = \dfrac{4}{256}.$

So we get: $\dfrac{1}{4} + \dfrac{1}{16} + \dfrac{1}{64} + \dfrac{1}{256} = \dfrac{64 + 16 + 4 + 1}{256} = \dfrac{85}{256}$

5. $\dfrac{1}{2}+\dfrac{1}{3}+\dfrac{1}{4}+\dfrac{1}{5}$

We can get first: $\dfrac{1}{2}+\dfrac{1}{4}=\dfrac{2}{4}+\dfrac{1}{4}=\dfrac{3}{4}$, and $\dfrac{1}{3}+\dfrac{1}{5}=\dfrac{8}{15}$.

So we get: $\dfrac{1}{2}+\dfrac{1}{3}+\dfrac{1}{4}+\dfrac{1}{5}=\dfrac{3}{4}+\dfrac{8}{15}=\dfrac{45+32}{60}=\dfrac{77}{60}$

6. $\dfrac{1}{10}+\dfrac{1}{100}+\dfrac{1}{1000}+\dfrac{1}{10000}$

1/10 = 1000/10000

1/100 = 100/10000

1/1000 = 10/10000

So we get: $\dfrac{1}{10}+\dfrac{1}{100}+\dfrac{1}{1000}+\dfrac{1}{10000}=\dfrac{1000+100+10+1}{10000}=\dfrac{1111}{10000}$

7. $\dfrac{1}{10}-\dfrac{1}{100}-\dfrac{1}{1000}-\dfrac{1}{10000}=\dfrac{1000-100-10-1}{10000}=\dfrac{1000-111}{10000}=\dfrac{889}{10000}$

8. $\dfrac{12}{20}+\dfrac{20}{30}+\dfrac{30}{42}+\dfrac{42}{56}$

To begin with, we get: 12/20 = 3/5, 20/30 = 2/3, 30/42 = 5/7, and 42/56 = 6/8 = 3/4.

Next, we can get: 3/5 = 1 – 2/5, 2/3 = 1 – 1/3, 5/7 = 1 – 2/7, and 3/4 = 1 – 1/4

So we get: 3/5 + 2/3 + 5/7 + 3/4 = 4 – (2/5 + 1/3 + 2/7 + 3/4)

Next, we can get: 2/5 + 1/3 = (6 + 5)/15 = 11/15, and 2/7 + 1/4 = (8 + 7)/28 = 15/28.

So next, we get:

$$\frac{11}{15}+\frac{15}{28}=\frac{11\cdot28}{15\cdot28}+\frac{15\cdot15}{15\cdot28}=\frac{280+28+150+75}{300+80+40}=\frac{308+225}{420}=\frac{533}{420}$$

And thus, we get:

$$\frac{12}{20}+\frac{20}{30}+\frac{30}{42}+\frac{42}{56}=4-\frac{533}{420}=4-\frac{420}{420}-\frac{113}{420}=3-\frac{113}{420}=2+\frac{420-113}{420}$$

$$=2+\frac{307}{420}=\frac{840+307}{420}=\frac{1147}{420}$$

9. $\dfrac{121}{11}+\dfrac{144}{12}+\dfrac{169}{13}+\dfrac{196}{14}$

To begin with, we have: $121 = 11^2$, $144 = 12^2$, $169 = 13^2$, and $196 = 14^2$.

So we get: $\dfrac{121}{11}+\dfrac{144}{12}+\dfrac{169}{13}+\dfrac{196}{14}=11+12+13+14=50.$

A. $\dfrac{1}{1}-\dfrac{1}{11}-\dfrac{11}{121}-\dfrac{1331}{121}-\dfrac{14641}{1331}$

First, we have: $121 = 11^2$, $1331 = 11 \times 121$, and $14641 = 1331 \times 11$.
So next, we can get: $11/121 = 1/11$, $1331/121 = 1/11$, and $14641/1331 = 1/11$
So we get:

$$\frac{1}{1}-\frac{1}{11}-\frac{11}{121}-\frac{1331}{121}-\frac{14641}{1331}=1-\frac{1}{11}-\frac{1}{11}-11-11=-\frac{2}{11}-21=-\frac{2+21\cdot11}{11}=-\frac{233}{11}$$

B. $\dfrac{55}{111}+\dfrac{333}{12321}+\dfrac{444}{24642}$

To begin with, we have: $12321 = 111^2$, and $24642 = 2\cdot 12321 = 2\cdot 111^2$.

So we get:

$$\frac{55}{111}+\frac{333}{12321}+\frac{444}{24642}=\frac{55}{111}+\frac{3\cdot 111}{111^2}+\frac{4\cdot 111}{2\cdot 111^2}=\frac{55}{111}+\frac{3}{111}+\frac{2}{111}=\frac{60}{111}=\frac{20}{37}$$

C. $\dfrac{2}{3}+\dfrac{3}{2}+\dfrac{3}{4}+\dfrac{4}{3}+\dfrac{4}{5}+\dfrac{5}{4}$

First, we can get: $2/3 + 4/3 = 6/3 = 2$, and $3/2 + 3/4 + 5/4 = (6 + 3 + 5)/4 = 14/4 = 7/2$.

So we get: $\dfrac{2}{3}+\dfrac{3}{2}+\dfrac{3}{4}+\dfrac{4}{3}+\dfrac{4}{5}+\dfrac{5}{4}=2+\dfrac{7}{2}+\dfrac{4}{5}=2+3+\dfrac{1}{2}+\dfrac{4}{5}=5+\dfrac{5+8}{10}=5+\dfrac{13}{10}=\dfrac{63}{10}$

D. $\dfrac{15}{12}+\dfrac{20}{15}+\dfrac{35}{28}+\dfrac{36}{27}$

To begin with, we can get: $15/12 = 5/4$, $20/15 = 4/3$, $35/28 = 5/4$, and $36/27 = 4/3$

So we get:

$$\frac{15}{12}+\frac{20}{15}+\frac{35}{28}+\frac{36}{27}=2\cdot\frac{5}{4}+2\cdot\frac{4}{3}=\frac{5}{2}+\frac{8}{3}=\frac{15+16}{6}=\frac{31}{6}.$$

Examples 4 in Rationals

0. $\dfrac{0.4}{0.5} - \dfrac{0.5}{0.6}$

1. $\dfrac{0.3}{0.4} - \dfrac{0.3}{0.17}$

2. $\dfrac{5}{6} + \dfrac{5}{12} + \dfrac{5}{20} + \dfrac{5}{30} + \dfrac{5}{42}$

3. $\dfrac{1}{15} + \dfrac{1}{35} + \dfrac{1}{63} + \dfrac{1}{99} + \dfrac{1}{143}$

4. $\dfrac{7}{10} + \dfrac{7}{40} + \dfrac{7}{88} + \dfrac{7}{154} + \dfrac{7}{238}$

Suggestions or Solutions
To the Examples 4 in Rationals

0. $\dfrac{0.4}{0.5}+\dfrac{0.5}{0.6}=\dfrac{4}{5}+\dfrac{5}{6}=\dfrac{24+25}{30}=\dfrac{49}{30}$

1. $\dfrac{0.3}{0.4}-\dfrac{0.3}{0.17}=\dfrac{3}{4}-\dfrac{30}{17}=\dfrac{51-120}{68}=\dfrac{-69}{68}$

We have: $\dfrac{a}{AB}=\dfrac{a}{B-A}(\dfrac{1}{A}-\dfrac{1}{B})$, where $A\neq 0$, $B\neq 0$, and $A\neq B$.

That's simply because: $\dfrac{a}{B-A}(\dfrac{1}{A}-\dfrac{1}{B})=\dfrac{a}{B-A}\cdot\dfrac{B-A}{AB}=\dfrac{a}{AB}$.

For instance, $\dfrac{2}{15}=\dfrac{2}{3\times 5}=\dfrac{2}{5-3}(\dfrac{1}{3}-\dfrac{1}{5})$. So using the fact above, we can get:

2. $\dfrac{5}{6}+\dfrac{5}{12}+\dfrac{5}{20}+\dfrac{5}{30}+\dfrac{5}{42}=\dfrac{5}{2\times 3}+\dfrac{5}{3\times 4}+\dfrac{5}{4\times 5}+\dfrac{5}{5\times 6}+\dfrac{5}{6\times 7}$

$=\dfrac{5}{3-2}(\dfrac{1}{2}-\dfrac{1}{3})+\dfrac{5}{4-3}(\dfrac{1}{3}-\dfrac{1}{4})+\dfrac{5}{5-4}(\dfrac{1}{4}-\dfrac{1}{5})+\dfrac{5}{6-5}(\dfrac{1}{5}-\dfrac{1}{6})+\dfrac{5}{7-6}(\dfrac{1}{6}-\dfrac{1}{7})$

$=5\cdot(\dfrac{1}{2}-\dfrac{1}{3})+5\cdot(\dfrac{1}{3}-\dfrac{1}{4})+5\cdot(\dfrac{1}{4}-\dfrac{1}{5})+5\cdot(\dfrac{1}{5}-\dfrac{1}{6})+5\cdot(\dfrac{1}{6}-\dfrac{1}{7})$

$=\dfrac{5}{2}-\dfrac{5}{3}+\dfrac{5}{3}-\dfrac{5}{4}+\dfrac{5}{4}-\dfrac{5}{5}+\dfrac{5}{5}-\dfrac{5}{6}+\dfrac{5}{6}-\dfrac{5}{7}=\dfrac{5}{2}-\dfrac{5}{7}=\dfrac{35-10}{14}=\dfrac{25}{14}$

3. $\dfrac{1}{15}+\dfrac{1}{35}+\dfrac{1}{63}+\dfrac{1}{99}+\dfrac{1}{143}=\dfrac{1}{3\times5}+\dfrac{1}{5\times7}+\dfrac{1}{7\times9}+\dfrac{1}{9\times11}+\dfrac{1}{11\times13}$

$=\dfrac{1}{5-3}(\dfrac{1}{3}-\dfrac{1}{5})+\dfrac{1}{7-5}(\dfrac{1}{5}-\dfrac{1}{7})+\dfrac{1}{9-7}(\dfrac{1}{7}-\dfrac{1}{9})+\dfrac{1}{11-9}(\dfrac{1}{9}-\dfrac{1}{11})+\dfrac{1}{13-11}(\dfrac{1}{11}-\dfrac{1}{13})$

$=\dfrac{1}{2}(\dfrac{1}{3}-\dfrac{1}{5})+\dfrac{1}{2}(\dfrac{1}{5}-\dfrac{1}{7})+\dfrac{1}{2}(\dfrac{1}{7}-\dfrac{1}{9})+\dfrac{1}{2}(\dfrac{1}{9}-\dfrac{1}{11})+\dfrac{1}{2}(\dfrac{1}{11}-\dfrac{1}{13})$

$=\dfrac{1}{2}\{(\dfrac{1}{3}-\dfrac{1}{5})+(\dfrac{1}{5}-\dfrac{1}{7})+(\dfrac{1}{7}-\dfrac{1}{9})+(\dfrac{1}{9}-\dfrac{1}{11})+(\dfrac{1}{11}-\dfrac{1}{13})\}$

$=\dfrac{1}{2}(\dfrac{1}{3}-\dfrac{1}{5}+\dfrac{1}{5}-\dfrac{1}{7}+\dfrac{1}{7}-\dfrac{1}{9}+\dfrac{1}{9}-\dfrac{1}{11}+\dfrac{1}{11}-\dfrac{1}{13})=\dfrac{1}{2}(\dfrac{1}{3}-\dfrac{1}{13})=\dfrac{1}{2}\cdot\dfrac{13-3}{39}=\dfrac{1}{2}\cdot\dfrac{10}{39}=\dfrac{5}{39}$

4. $\dfrac{7}{10}+\dfrac{7}{40}+\dfrac{7}{88}+\dfrac{7}{154}+\dfrac{7}{238}=7(\dfrac{1}{2\times5}+\dfrac{1}{5\times8}+\dfrac{1}{8\times11}+\dfrac{1}{11\times14}+\dfrac{1}{14\times17})$

$=\dfrac{7}{5-2}(\dfrac{1}{2}-\dfrac{1}{5})+\dfrac{7}{8-5}(\dfrac{1}{5}-\dfrac{1}{8})+\dfrac{7}{11-8}(\dfrac{1}{8}-\dfrac{1}{11})+\dfrac{7}{14-11}(\dfrac{1}{11}-\dfrac{1}{14})+\dfrac{7}{17-14}(\dfrac{1}{14}-\dfrac{1}{17})$

$=\dfrac{7}{3}(\dfrac{1}{2}-\dfrac{1}{5})+\dfrac{7}{3}(\dfrac{1}{5}-\dfrac{1}{8})+\dfrac{7}{3}(\dfrac{1}{8}-\dfrac{1}{11})+\dfrac{7}{3}(\dfrac{1}{11}-\dfrac{1}{14})+\dfrac{7}{3}(\dfrac{1}{14}-\dfrac{1}{17})$

$=\dfrac{7}{3}\{(\dfrac{1}{2}-\dfrac{1}{5})+(\dfrac{1}{5}-\dfrac{1}{8})+(\dfrac{1}{8}-\dfrac{1}{11})+(\dfrac{1}{11}-\dfrac{1}{14})+(\dfrac{1}{14}-\dfrac{1}{17})\}$

$=\dfrac{7}{3}(\dfrac{1}{2}-\dfrac{1}{5}+\dfrac{1}{5}-\dfrac{1}{8}+\dfrac{1}{8}-\dfrac{1}{11}+\dfrac{1}{11}-\dfrac{1}{14}+\dfrac{1}{14}-\dfrac{1}{17})$

$=\dfrac{7}{3}(\dfrac{1}{2}-\dfrac{1}{17})=\dfrac{7}{3}\cdot\dfrac{17-2}{34}=\dfrac{7}{3}\cdot\dfrac{15}{34}=\dfrac{7\cdot5}{34}=\dfrac{35}{34}$

3. Rational Numbers 3

First, going over some ideas in the previous sections, we can begin with:

- $$\frac{A}{B} = \frac{AC}{BC} = \frac{AE}{BE} = \frac{AXY}{BXY} = \ldots, \text{ where } B, C, D, E, X, \text{ and } Y \neq 0.$$

So other than 0, multiplying by the same both the numerator and the denominator, we get the same ratio, that is, the value of the fraction (or the ratio) does not change.

And because of the fact above, adding fractions, we do the addition the way below:

- $$\frac{A}{B} + \frac{C}{D} = \frac{AD}{BD} + \frac{BC}{BD} = \frac{AD + BC}{BD}, \text{ where } B \text{ and } D \neq 0.$$

For instance: $\dfrac{2}{3} + \dfrac{5}{7} = \dfrac{2 \times 7}{3 \times 7} + \dfrac{5 \times 3}{7 \times 3} = \dfrac{14}{21} + \dfrac{15}{21} = \dfrac{14 + 15}{21} = \dfrac{29}{21}.$

And next, moving on to subtractions, we can do a subtraction the way below:

- $$\frac{A}{B} - \frac{C}{D} = \frac{AD}{BD} - \frac{BC}{BD} = \frac{AD - BC}{BD}, \text{ where } B \text{ and } D \neq 0.$$

For instance: $\dfrac{2}{3} - \dfrac{5}{7} = \dfrac{2 \times 7}{3 \times 7} - \dfrac{5 \times 3}{7 \times 3} = \dfrac{14}{21} - \dfrac{15}{21} = \dfrac{14 - 15}{21} = \dfrac{-1}{21} = -\dfrac{1}{21}.$

So in sum, we have: $\dfrac{A}{B} \pm \dfrac{C}{D} = \dfrac{AD}{BD} \pm \dfrac{BC}{BD} = \dfrac{AD \pm BC}{BD}, \text{ where } B \text{ and } D \neq 0.$

And next, moving on to multiplications, we can say that taking a product of fractions, we get a fraction where the numerator is the product of the numerators, and the denominator is the product of the denominators of the fractions multiplied.

In short, we get: A/B x $C/D = AC/BD$, that is, we get: $\dfrac{A}{B}$ x $\dfrac{C}{D} = \dfrac{AC}{BD}$.

Now, moving on to divisions, and dividing, for instance, 1 by 1/3, what do we get?

We get 3, because 1 is three times 1/3. That is, three of (1/3)s make 1.

So we get: $1 \div 1$ x $3 = 3$. That is, we get: $1 \div \frac{1}{3} = 3$. And we have: $\frac{1}{3}$ x $3 = 3$ x $\frac{1}{3} = 1$.

What then, do we get if we divide 2 by 1/3?

We know 2 is two of 1s.

So we can divide 2 by 1/3 doing twice the division of 1 by 1/3, and then, adding together the two results, which are two of 3s. Then, we get 6.

And putting the idea above in expressions, we can put it the way below:
$(1 \div \frac{1}{3}) + (1 \div \frac{1}{3}) = 2$ x $(1 \div \frac{1}{3}) = 2$ x $3 = 6$.

And we can put the same the way below, too:

Putting one pizza on top of another same pizza, we get a stack of two same pizzas. Then, dividing the stack by 3, what do we get?

We get six equal pieces of pizza, and every piece is 1/3 of one pizza.
So we get: $2 \div \frac{1}{3} = 6$.

And we have this, too: $(1 \div \frac{1}{3}) + (1 \div \frac{1}{3}) = 2 \text{ x } (1 \div \frac{1}{3}) = 2 \text{ x } 3 = 6$.

And we know: $2 \div \frac{1}{3} = (1 \div \frac{1}{3}) + (1 \div \frac{1}{3}) = 2 \text{ x } 3$.

So we get: $2 \div \frac{1}{3} = 2 \text{ x } 3 = 6$. (Notice that dividing by 1/3, we can multiply by 3.)

And of course, we can put the result above the way below, too:

Since 1 is three times 1/3, and 2 is twice 1, we can say that 2 is six times 1/3.

Moving next, on to another example, and dividing 4/5 by 2/3, what do we get?

We have: 4/5 = 12/15, and 2/3 = 10/15.

So dividing 4/5 by 2/3, we can divide 12/15 by 10/15.

And we know 12/15 is 12 of (1/15)s, and 10/15 is 10 of (1/15)s.

And thus, dividing 4/5 by 2/3, we in fact, divide 12 by 10, that is, we get: 12/10 = 6/5.

So we get: $\frac{4}{5} \div \frac{2}{3} = \frac{12}{15} \div \frac{10}{15} = 12 \div 10 = \frac{12}{10} = \frac{6}{5}$.

Now, how then, do we get 12 and 10 in the calculation above?

We get 12 multiplying 4 by 3, and get 10 multiplying 5 by 2. Specifically:

• Multiplying the numerator of 4/5 by the denominator of 2/3, we get 12, which is the numerator of 12/10, which is the result.

• Multiplying the denominator of 4/5 by the numerator of 2/3, we get 10, which is the denominator of 12/10.

So we can notice that dividing a fraction by another, we can multiply the fraction by the reciprocal of the other. What reciprocal though?

For instance, the reciprocal of 2/3 is 3/2, and the reciprocal of 3/2 is 2/3.

So the product of two numbers reciprocal to each other is 1.

So we can put the division above this way: $\frac{4}{5} \div \frac{2}{3} = \frac{4}{5} \times \frac{3}{2} = \frac{12}{10} = \frac{6}{5}$.

In short, we get: $\frac{A}{B} \div \frac{C}{D} = \frac{A}{B} \times \frac{D}{C} = \frac{AD}{BC}$, where B, C, and $D \neq 0$, of course.

And now, putting threads altogether, we have:

$$\frac{A}{B} \pm \frac{C}{D} = \frac{AD}{BD} \pm \frac{BC}{BD} = \frac{AD \pm BC}{BD}, \text{ where } B \text{ and } D \neq 0.$$

$$\frac{A}{B} \times \frac{C}{D} = \frac{A}{B} \cdot \frac{C}{D} = \frac{A\ C}{B\ D} = \frac{AC}{BD}, \text{ where } B \text{ and } D \neq 0.$$

$$\frac{A}{B} \div \frac{C}{D} = \frac{A}{B} \times \frac{D}{C} = \frac{AD}{BC}, \text{ where } B, C, \text{ and } D \neq 0.$$

And in fact, we can use any real number in the operations above if it is appropriate, that is, if it does not make any denominator 0.

For instance, we can have:

$$\frac{0.31}{0.2} + \frac{1.7}{2.5} = \frac{0.31 \times 2.5}{0.2 \times 2.5} + \frac{0.2 \times 1.7}{0.2 \times 2.5} = \frac{(0.31 \times 2.5) + (0.2 \times 1.7)}{0.2 \times 2.5} = \frac{0.775 + 0.34}{0.2 \times 2.5} = \frac{1.115}{0.5}.$$

$$\frac{0.31}{0.2} - \frac{1.7}{2.5} = \frac{0.31 \times 2.5}{0.2 \times 2.5} - \frac{0.2 \times 1.7}{0.2 \times 2.5} = \frac{(0.31 \times 2.5) - (0.2 \times 1.7)}{0.2 \times 2.5} = \frac{0.775 - 0.34}{0.2 \times 2.5} = \frac{0.435}{0.5}.$$

$$\frac{0.31}{0.2} \times \frac{1.7}{2.5} = \frac{0.31}{0.2} \cdot \frac{1.7}{2.5} = \frac{0.31 \times 1.7}{0.2 \times 2.5} = \frac{0.527}{0.5} = \frac{5.27}{5} = \frac{527}{500}.$$

$$\frac{0.31}{0.2} \div \frac{1.7}{2.5} = \frac{0.31}{0.2} \times \frac{2.5}{1.7} = \frac{0.31 \times 2.5}{0.2 \times 1.7} = \frac{0.775}{0.34} = \frac{775}{340} = \frac{155}{68}.$$

$$\frac{\frac{2}{3}}{\frac{5}{7}} + \frac{\frac{9}{5}}{\frac{3}{8}} = \frac{\frac{2}{3} \times \frac{3}{8}}{\frac{5}{7} \times \frac{3}{8}} + \frac{\frac{5}{7} \times \frac{9}{5}}{\frac{5}{7} \times \frac{3}{8}} = \frac{\left(\frac{2}{3} \times \frac{3}{8}\right) + \left(\frac{5}{7} \times \frac{9}{5}\right)}{\frac{5}{7} \times \frac{3}{8}} = \frac{\frac{6}{24} + \frac{45}{35}}{\frac{15}{56}} = \frac{\frac{6 \times 35 + 24 \times 45}{24 \times 35}}{\frac{15}{56}} = \frac{\frac{210 + 1080}{840}}{\frac{15}{56}} = \frac{\frac{1290}{840}}{\frac{15}{56}}$$

$$= \frac{\frac{129}{84}}{\frac{15}{56}} = \frac{\frac{129}{84} \times 56}{\frac{15}{56} \times 56} = \frac{\frac{7224}{84}}{15} = \frac{\frac{7224}{84} \times 84}{15 \times 84} = \frac{7224}{1260} = \frac{1806 \times 4}{315 \times 4} = \frac{1086}{315} = \frac{602 \times 3}{105 \times 3} = \frac{602}{105} = \frac{86}{15}.$$

So we can calculate $\dfrac{\frac{2}{3}}{\frac{5}{7}} + \dfrac{\frac{9}{5}}{\frac{3}{8}}$ the way below:

First, we can have: $\dfrac{\frac{2}{3}}{\frac{5}{7}} = \dfrac{\frac{2}{3} \times 7}{\frac{5}{7} \times 7} = \dfrac{\frac{14}{3}}{5} = \dfrac{\frac{14}{3} \times 3}{5 \times 3} = \dfrac{14}{15}$, and $\dfrac{\frac{9}{5}}{\frac{3}{8}} = \dfrac{\frac{9}{5} \times 8}{\frac{3}{8} \times 8} = \dfrac{\frac{72}{5}}{3} = \dfrac{\frac{72}{5} \times 5}{3 \times 5} = \dfrac{72}{15}.$

So next, we get: $\dfrac{\frac{2}{3}}{\frac{5}{7}} + \dfrac{\frac{9}{5}}{\frac{3}{8}} = \dfrac{14}{15} + \dfrac{72}{15} = \dfrac{86}{15}.$ And we can get the sum faster the way below:

To begin with, we can have: $\dfrac{\frac{A}{B}}{\frac{C}{D}} = \dfrac{\frac{A}{B}D}{\frac{C}{D}D} = \dfrac{\frac{AD}{B}}{C} = \dfrac{\frac{AD}{B}\frac{1}{C}}{C\frac{1}{C}} = \dfrac{\frac{AD}{BC}}{1} = \dfrac{AD}{BC}.$

And we can put it the way below, too:

We know we can get: $\dfrac{1}{\frac{C}{D}} = \dfrac{1 \cdot D}{\frac{C}{D}D} = \dfrac{D}{C}.$ So we can get: $\dfrac{\frac{A}{B}}{\frac{C}{D}} = \dfrac{A}{B} \cdot \dfrac{1}{\frac{C}{D}} = \dfrac{A}{B}\dfrac{D}{C} = \dfrac{AD}{BC}.$

So either way, we get: $\dfrac{\frac{A}{B}}{\frac{C}{D}} = \dfrac{AD}{BC}.$

And thus, calculating $\dfrac{\frac{2}{3}}{\frac{5}{7}} + \dfrac{\frac{9}{5}}{\frac{3}{8}}$, we can get the sum faster the way below:

$$\frac{\frac{2}{3}}{\frac{5}{7}}+\frac{\frac{9}{5}}{\frac{3}{8}}=\frac{2\cdot7}{3\cdot5}+\frac{9\cdot8}{5\cdot3}=\frac{14+72}{15}=\frac{86}{15}.$$

And by the same token, we can calculate $\dfrac{\frac{2}{3}}{\frac{5}{7}}-\dfrac{\frac{9}{5}}{\frac{3}{8}}$ the way below:

$$\frac{\frac{2}{3}}{\frac{5}{7}}-\frac{\frac{9}{5}}{\frac{3}{8}}=\frac{2\cdot7}{3\cdot5}-\frac{9\cdot8}{5\cdot3}=\frac{14-72}{15}=-\frac{58}{15}.$$

Taking some more instances, we can have:

$$\frac{-\frac{2}{5}}{\frac{7}{9}}-\frac{\frac{3}{4}}{-\frac{5}{3}}=-\frac{\frac{2}{5}}{\frac{7}{9}}+\frac{\frac{3}{4}}{\frac{5}{3}}=-\frac{2\cdot9}{5\cdot7}+\frac{3\cdot3}{4\cdot5}=\frac{-2\cdot9\cdot4+3\cdot3\cdot7}{4\cdot5\cdot7}=\frac{-72+63}{140}=-\frac{9}{140}.$$

$$\frac{\frac{12}{25}}{\frac{8}{15}}=\frac{\frac{3}{25}}{\frac{2}{15}}=\frac{\frac{3}{5}}{\frac{2}{3}}=\frac{9}{10}.$$

$$\frac{\frac{12}{5}}{\frac{9}{25}}=\frac{\frac{4}{5}}{\frac{3}{25}}=\frac{\frac{4}{1}}{\frac{3}{5}}=\frac{20}{3}.$$

$$\frac{\frac{12}{15}}{\frac{9}{15}}=\frac{\frac{12}{1}}{\frac{9}{1}}=\frac{12}{9}=\frac{4}{3}.$$

$$\frac{\frac{12}{15}}{\frac{9}{15}}=\frac{\frac{12}{1}}{\frac{9}{1}}=\frac{\frac{4}{1}}{\frac{3}{1}}=\frac{4}{3}.$$

$$\frac{\frac{4}{3}}{\frac{8}{9}}=\frac{\frac{1}{3}}{\frac{2}{9}}=\frac{\frac{1}{1}}{\frac{2}{3}}=\frac{1}{\frac{2}{3}}=\frac{3}{2}.$$

$$\frac{\frac{16}{15}}{\frac{8}{24}}=\frac{\frac{16}{5}}{\frac{8}{8}}=\frac{\frac{16}{5}}{1}=\frac{16}{5}.$$

$$\frac{\frac{a+b}{c}}{\frac{p-q}{r}}=\frac{(a+b)r}{(p-q)c}.$$

Examples 5 in Rationals

0. $\dfrac{3}{4} \times \dfrac{7}{5}$

1. $\dfrac{12}{32} \times \dfrac{18}{15}$

2. $\dfrac{4}{5} \times \dfrac{5}{4}$

3. $\dfrac{2}{3} \times \dfrac{5}{6} \times \dfrac{3}{7} \times \dfrac{6}{15} \times \dfrac{14}{3} \times \dfrac{3}{2}$

4. $\dfrac{3}{4} \div 7$

5. $\dfrac{5}{6} \div \dfrac{10}{3}$

6. $\dfrac{12}{5} \div \dfrac{3}{10}$

7. $3.5 \div \dfrac{14}{5}$

8. $\dfrac{4}{9} \div \dfrac{7}{12} \div \dfrac{16}{3} \div \dfrac{14}{9}$

9. $\dfrac{\dfrac{1}{4} \div [62 + \dfrac{1}{4} - \{8.5 - 0.4 \times (1 + \dfrac{7}{8})\} \div 0.125]}{[2 - \{5.55 \times (1 + \dfrac{1}{3}) - 2.7 \div 0.4\}] \div 0.135}$

A. $\dfrac{\dfrac{7}{18} \times (4 + \dfrac{1}{2}) \times \dfrac{1}{6}}{\{(13 + \dfrac{1}{3}) - (3 + \dfrac{3}{4}) \div \dfrac{5}{16}\} \times (2 + \dfrac{7}{8})}$

B. $\dfrac{(\dfrac{1}{3} \div 0.3) \times 0.1}{0.8 \times (1.5 - \dfrac{1}{4})}$

C. $(2 + \dfrac{2}{5}) - \dfrac{1}{\frac{1}{2} + \frac{1}{1 - \frac{1}{2}}}$

D. $\dfrac{0.5 + 4.8 \times \dfrac{1}{6}}{(0.5 \times 4.8 \times \dfrac{1}{6}) + 1 + \dfrac{1}{1 - \dfrac{1}{2}}}$

E. $\left[36.6 - \left[\left(\dfrac{6}{3 + \frac{1}{3}} \right) + \dfrac{1}{3} \right] \times (4 + \dfrac{1}{2}) \right] \div \left[(7 + \dfrac{1}{20}) + 6.35 \right]$

Suggestions or Solutions
To the Examples 5 in Rationals

0. $\dfrac{3}{4} \times \dfrac{7}{5} = \dfrac{3 \times 7}{4 \times 5} = \dfrac{21}{20}$

1. $\dfrac{12}{32} \times \dfrac{18}{15} = \dfrac{3 \times 4}{8 \times 4} \times \dfrac{6 \times 3}{5 \times 3} = \dfrac{3}{8} \times \dfrac{6}{5} = \dfrac{3 \times 6}{8 \times 5} = \dfrac{18}{40} = \dfrac{9}{20}$

2. $\dfrac{4}{5} \times \dfrac{5}{4} = \dfrac{20}{20} = 1$

3. $\dfrac{2}{3} \times \dfrac{5}{6} \times \dfrac{3}{7} \times \dfrac{6}{15} \times \dfrac{14}{3} \times \dfrac{3}{2} = \dfrac{2}{3} \times \dfrac{3}{7} \times \dfrac{5}{6} \times \dfrac{6}{15} \times \dfrac{14}{3} \times \dfrac{3}{2}$

$= (\dfrac{2}{3} \times \dfrac{3}{7}) \times (\dfrac{5}{6} \times \dfrac{6}{15}) \times (\dfrac{14}{3} \times \dfrac{3}{2}) = \dfrac{2}{7} \times \dfrac{5}{15} \times \dfrac{14}{2} = \dfrac{2}{7} \times \dfrac{1}{3} \times 7 = \dfrac{2}{7} \times 7 \times \dfrac{1}{3} = 2 \times \dfrac{1}{3} = \dfrac{2}{3}$

4. $\dfrac{3}{4} \div 7 = \dfrac{3}{4} \times \dfrac{1}{7} = \dfrac{3}{28}$

5. $\dfrac{5}{6} \div \dfrac{10}{3} = \dfrac{5}{6} \times \dfrac{3}{10} = \dfrac{5}{2 \times 3} \times \dfrac{3}{5 \times 2} == \dfrac{1}{2} \times \dfrac{1}{2} = \dfrac{1}{4}$

6. $\dfrac{12}{5} \div \dfrac{3}{10} = \dfrac{12}{5} \times \dfrac{10}{3} = \dfrac{4 \times 3}{5} \times \dfrac{5 \times 2}{3} = \dfrac{4 \times 3}{1} \times \dfrac{1 \times 2}{3} = \dfrac{4 \times 1}{1} \times \dfrac{1 \times 2}{1} = 4 \times 2 = 8$

7. $3.5 \div \dfrac{14}{5} = \dfrac{7}{2} \times \dfrac{5}{14} = \dfrac{7}{2} \times \dfrac{5}{7 \times 2} = \dfrac{1}{2} \times \dfrac{5}{1 \times 2} = \dfrac{5}{4}$

8. $\dfrac{4}{9} \div \dfrac{7}{12} \div \dfrac{16}{3} \div \dfrac{14}{9} = \dfrac{4}{9} \times \dfrac{12}{7} \times \dfrac{3}{16} \times \dfrac{9}{14} = \dfrac{4}{3} \times \dfrac{4}{7} \times \dfrac{3}{16} \times \dfrac{9}{14}$

$= \dfrac{4}{1} \times \dfrac{4}{7} \times \dfrac{1}{16} \times \dfrac{9}{14} = \dfrac{4}{1} \times \dfrac{1}{7} \times \dfrac{1}{4} \times \dfrac{9}{14} = \dfrac{1}{1} \times \dfrac{1}{7} \times \dfrac{1}{1} \times \dfrac{9}{14} = \dfrac{1}{7} \times \dfrac{9}{14} = \dfrac{9}{98}$

9.
$$\dfrac{\dfrac{1}{4} \div [62 + \dfrac{1}{4} - \{8.5 - 0.4 \times (1 + \dfrac{7}{8})\} \div 0.125]}{[2 - \{5.55 \times (1 + \dfrac{1}{3}) - 2.7 \div 0.4\}] \div 0.135}$$

$$= \dfrac{\dfrac{1}{4} \div [62 + \dfrac{1}{4} - \left(\dfrac{85}{10} - \dfrac{4}{10} \times \dfrac{15}{8}\right) \div \dfrac{125}{1000}]}{\left[2 - \left(\dfrac{555}{100} \times \dfrac{4}{3} - \dfrac{27}{10} \div \dfrac{4}{10}\right)\right] \div \dfrac{135}{1000}} = \dfrac{\dfrac{1}{4} \div [62 + \dfrac{1}{4} - \left(\dfrac{17 \times 5}{2 \times 5} - \dfrac{2 \times 2}{2 \times 5} \times \dfrac{15}{8}\right) \div \dfrac{125}{125 \times 8}]}{\left[2 - \left(\dfrac{111 \times 5}{20 \times 5} \times \dfrac{4}{3} - \dfrac{27}{10} \div \dfrac{2 \times 2}{2 \times 5}\right)\right] \div \dfrac{27 \times 5}{200 \times 5}}$$

$$= \dfrac{\dfrac{1}{4} \div [62 + \dfrac{1}{4} - \left(\dfrac{17}{2} - \dfrac{2}{5} \times \dfrac{15}{8}\right) \div \dfrac{1}{8}]}{\left[2 - \left(\dfrac{111}{20} \times \dfrac{4}{3} - \dfrac{27}{10} \div \dfrac{2}{5}\right)\right] \div \dfrac{27}{200}} = \dfrac{\dfrac{1}{4} \div [62 + \dfrac{1}{4} - \left(\dfrac{17}{2} - \dfrac{1}{1} \times \dfrac{3}{4}\right) \times 8]}{\left[2 - \left(\dfrac{111}{20} \times \dfrac{4}{3} - \dfrac{27}{10} \times \dfrac{5}{2}\right)\right] \times \dfrac{200}{27}}$$

$$= \dfrac{\dfrac{1}{4} \div [62 + \dfrac{1}{4} - \left(\dfrac{17}{2} - \dfrac{3}{4}\right) \times 8]}{\left[2 - \left(\dfrac{37}{5} \times \dfrac{1}{1} - \dfrac{27}{2} \times \dfrac{1}{2}\right)\right] \times \dfrac{200}{27}} = \dfrac{\dfrac{1}{4} \div [62 + \dfrac{1}{4} - \left(\dfrac{34}{4} - \dfrac{3}{4}\right) \times 8]}{\left[2 - \left(\dfrac{37}{5} - \dfrac{27}{4}\right)\right] \times \dfrac{200}{27}}$$

$$= \dfrac{\dfrac{1}{4} \div [62 + \dfrac{1}{4} - \dfrac{31}{4} \times 8]}{\left[2 - \dfrac{148 - 135}{20}\right] \times \dfrac{200}{27}} = \dfrac{\dfrac{1}{4} \div [62 + \dfrac{1}{4} - 31 \times 2]}{\left[2 - \dfrac{13}{20}\right] \times \dfrac{200}{27}} = \dfrac{\dfrac{1}{4} \div [62 + \dfrac{1}{4} - 62]}{\dfrac{27}{20} \times \dfrac{200}{27}} = \dfrac{\dfrac{1}{4} \div \dfrac{1}{4}}{\dfrac{1}{1} \times \dfrac{10}{1}} = \dfrac{1}{10}$$

A. $\dfrac{\dfrac{7}{18} \times (4+\dfrac{1}{2}) \times \dfrac{1}{6}}{\{(13+\dfrac{1}{3})-(3+\dfrac{3}{4})\div\dfrac{5}{16}\} \times (2+\dfrac{7}{8})} = \dfrac{\dfrac{7}{18} \times \dfrac{9}{2} \times \dfrac{1}{6}}{(\dfrac{40}{3}-\dfrac{15}{4} \times \dfrac{16}{5}) \times \dfrac{23}{8}} = \dfrac{\dfrac{7}{2} \times \dfrac{1}{2} \times \dfrac{1}{6}}{(\dfrac{40}{3}-\dfrac{3}{1} \times \dfrac{4}{1}) \times \dfrac{23}{8}}$

$= \dfrac{\dfrac{7}{4} \times \dfrac{1}{6}}{(\dfrac{40}{3}-12) \times \dfrac{23}{8}} = \dfrac{\dfrac{7}{24}}{\dfrac{4}{3} \times \dfrac{23}{8}} = \dfrac{\dfrac{7}{24}}{\dfrac{1}{3} \times \dfrac{23}{2}} = \dfrac{\dfrac{7}{24}}{\dfrac{23}{6}} = \dfrac{\dfrac{7}{24} \times 6}{\dfrac{23}{6} \times 6} = \dfrac{\dfrac{7}{4}}{23} = \dfrac{\dfrac{7}{4} \times 4}{23 \times 4} = \dfrac{7}{92}$

B. $\dfrac{(\dfrac{1}{3}\div0.3) \times 0.1}{0.8 \times (1.5-\dfrac{1}{4})} = \dfrac{(\dfrac{1}{3}\div\dfrac{3}{10}) \times \dfrac{1}{10}}{\dfrac{8}{10} \times (\dfrac{15}{10}-\dfrac{1}{4})} = \dfrac{\dfrac{1}{3} \times \dfrac{10}{3} \times \dfrac{1}{10}}{\dfrac{8}{10} \times \dfrac{30-5}{20}} = \dfrac{\dfrac{1}{3} \times \dfrac{1}{3} \times \dfrac{1}{1}}{\dfrac{8}{10} \times \dfrac{25}{20}} = \dfrac{\dfrac{1}{9}}{\dfrac{8}{10} \times \dfrac{5}{4}}$

$= \dfrac{\dfrac{1}{9}}{\dfrac{2}{2} \times \dfrac{1}{1}} = \dfrac{\dfrac{1}{9}}{1 \times 1} = \dfrac{1}{9}$

C. $(2+\dfrac{2}{5}) - \dfrac{1}{\dfrac{1}{2}+\dfrac{1}{1-\frac{1}{2}}}$

To begin with, we can have: $\dfrac{1}{\dfrac{1}{2}+\dfrac{1}{1-\frac{1}{2}}} = \dfrac{1}{\dfrac{1}{2}+\dfrac{1}{\frac{1}{2}}} = \dfrac{1}{\dfrac{1}{2}+\dfrac{1\times2}{\frac{1}{2}\times2}} = \dfrac{1}{\dfrac{1}{2}+\dfrac{2}{1}} = \dfrac{1}{\dfrac{1}{2}+2} = \dfrac{1}{\frac{5}{2}} = \dfrac{2}{5}$.

So we get: $(2+\dfrac{2}{5}) - \dfrac{1}{\dfrac{1}{2}+\dfrac{1}{1-\frac{1}{2}}} = (2+\dfrac{2}{5})-\dfrac{2}{5} = 2+\dfrac{2}{5}-\dfrac{2}{5} = 2.$

D.
$$\frac{0.5+4.8 \times \dfrac{1}{6}}{(0.5 \times 4.8 \times \dfrac{1}{6})+1+\dfrac{1}{1-\dfrac{1}{2}}} = \frac{\dfrac{1}{2}+\dfrac{48}{10}\times\dfrac{1}{6}}{(\dfrac{1}{2}+\dfrac{48}{10}\times\dfrac{1}{6})+1+\dfrac{1}{\dfrac{1}{2}}} = \frac{\dfrac{1}{2}+\dfrac{8}{10}\times\dfrac{1}{1}}{(\dfrac{1}{2}+\dfrac{8}{10}\times\dfrac{1}{1})+1+2}$$

$$= \frac{\dfrac{1}{1}+\dfrac{4}{10}}{(\dfrac{1}{1}+\dfrac{4}{10})+3} = \frac{1+\dfrac{2}{5}}{1+\dfrac{2}{5}+3} = \frac{\dfrac{7}{5}}{4+\dfrac{2}{5}} = \frac{\dfrac{7}{5}}{\dfrac{22}{5}} = \frac{\dfrac{7}{5}\times 5}{\dfrac{22}{5}\times 5} = \frac{7}{22}$$

E.
$$\left[36.6-\left[\left(\frac{6}{3+\frac{1}{3}}\right)+\frac{1}{3}\right] \times (4+\frac{1}{2})\right] \div \left[(7+\frac{1}{20})+6.35\right]$$

$$=\left[\frac{366}{10}-\left[\left(\frac{6}{\frac{10}{3}}\right)+\frac{1}{3}\right]\times\frac{9}{2}\right]\div\left[\frac{141}{20}+\frac{635}{100}\right]=\left[\frac{366}{10}-\left[\frac{18}{10}+\frac{1}{3}\right]\times\frac{9}{2}\right]\div\left[\frac{705}{100}+\frac{635}{100}\right]$$

$$=\left[\frac{366}{10}-\left[\frac{54+10}{30}\right]\times\frac{9}{2}\right]\div\left[\frac{1340}{100}\right]=\left[\frac{366}{10}-\frac{64}{30}\times\frac{9}{2}\right]\div\frac{134}{10}$$

$$=\left[\frac{366}{10}-\frac{32}{10}\times\frac{3}{1}\right]\times\frac{10}{134}=\left[\frac{366}{10}-\frac{96}{10}\right]\times\frac{10}{134}=\frac{270}{10}\times\frac{10}{134}=\frac{270}{134}=\frac{135}{67}$$

Examples 6 in Rationals

0. $$\frac{1+2+3+4+5+6+7+8+7+6+5+4+3+2+1}{88888888 \times 88888888}$$

1. $$1-\frac{1}{10}-\frac{1}{100}-\frac{1}{1000}-\text{.... ....}-\frac{1}{10000000000}$$

2. $$\frac{1234567890}{1234567891^2 - 1234567890 \times 1234567892}$$

3. $$(1+\frac{1}{2}) \times (1-\frac{1}{2}) \times (1+\frac{1}{3}) \times (1-\frac{1}{3}) \times (1+\frac{1}{100}) \times (1-\frac{1}{100})$$

48

Suggestions or Solutions
To the Examples 6 in Rationals

0. $\dfrac{1+2+3+4+5+6+7+8+7+6+5+4+3+2+1}{88888888 \times 88888888}$

To begin with, we can have:

$1 + 2 + 3 + 4 + 5 + 6 + 7 + 8 +$
$\underline{7 + 6 + 5 + 4 + 3 + 2 + 1}$
$8 + 8 + 8 + 8 + 8 + 8 + 8 + 8 = 8 \times 8$

So we get:

$$\frac{1+2+3+4+5+6+7+8+7+6+5+4+3+2+1}{88888888 \times 88888888} = \frac{8 \times 8}{88888888 \times 88888888}$$

$$= \frac{8 \times 8}{8 \times 11111111 \times 8 \times 11111111} = \frac{8 \times 8}{8 \times 8 \times 11111111^2} = \frac{1}{11111111^2}$$

And we have:
$11^2 = 121$
$111^2 = 12321$
$1111^2 = 1234321$

So by the same token, we can get: $11111111^2 = 123456787654321$

Thus, we get: $\dfrac{1+2+3+4+5+6+7+8+7+6+5+4+3+2+1}{88888888 \times 88888888} = \dfrac{1}{123456787654321}.$

1. $1 - \dfrac{1}{10} - \dfrac{1}{100} - \dfrac{1}{1000} - .... - \dfrac{1}{10000000000}$

To begin with, we know: $\dfrac{1}{10} = 0.1$, $\dfrac{1}{100} = 0.01$, $\dfrac{1}{1000} = 0.001$, and so forth.

So we get:

$1 - \dfrac{1}{10} - \dfrac{1}{100} - \dfrac{1}{1000} - .... - \dfrac{1}{10000000000}$

$= 1 - (0.1 + 0.01 + 0.001 + ... + 0.0000000001)$

$= 1 - 0.1111111111$

$= 0.8888888889$

2. $\dfrac{1234567890}{1234567891^2 - 1234567890 \text{ x } 1234567892}$

Assuming first, 1234567890 = A, we can get:

1234567891 = A + 1, and 1234567892 = A + 2.

So next, we get:

$$\dfrac{1234567890}{1234567891^2 - 1234567890 \text{ x } 1234567892} = \dfrac{A}{(A+1)^2 - A(A+2)}$$

$$= \dfrac{A}{A^2 + 2A + 1 - A^2 - 2A} = \dfrac{A}{1} = A$$

Thus, we get: $\dfrac{1234567890}{1234567891^2 - 1234567890 \text{ x } 1234567892} = 1234567890.$

3. $(1+\frac{1}{2}) \times (1-\frac{1}{2}) \times (1+\frac{1}{3}) \times (1-\frac{1}{3}) \times (1+\frac{1}{100}) \times (1-\frac{1}{100})$

$= \frac{3}{2} \times \frac{1}{2} \times \frac{4}{3} \times \frac{2}{3} \times \frac{101}{100} \times \frac{99}{100} = \frac{1}{1} \times \frac{1}{1} \times \frac{1}{1} \times \frac{2}{1} \times \frac{101}{100} \times \frac{33}{100} = \frac{101}{100} \times \frac{66}{100}$

$= \frac{6666}{10000} = \frac{3333}{5000}$

Examples 7 in Rationals

0.   Convert the following decimals into the fractional numbers.

0.0.   0.375375375…

0.1.   0.476547654765…

0.2.   2.0496496496…

0.3.   7.2060706070607…

1.   Some fractional numbers can be put in decimals. For instance:

1/7 = 0.142857142857142857…

2/7 = 0.285714285714285714…

3/7 = 0.428571428571428571…

1.0.   Find the 17th digit below the point in the decimal equivalent of 8/7.

1.1.   Find the 24th digit below the point in the decimal equivalent of 30/7.

2.   Convert the following fractional numbers into their decimal equivalents, and find the thousandth digit below the point in each of the decimals.

2.0.   5/33

2.1.   29/222

3.   Find the thousandth digit below the point in the product of the two decimals in each case below. Use a calculator.

3.0.   0.4122412241224… and 0.9443594435944359…

3.1.   1.124545454… and 4.2879269269269…

Suggestions or Solutions
To the Examples 7 in Rationals

0. Convert each decimal into its fractional equivalent.

0.0.   0.375375375…

We can see some digits repeating, and they are 375.

So assuming: $A = 0.375375375…$, we can set: $1000A = 375.375375375…$.

That is, we get: $1000A = 375 + 0.375375375…$.

And we know: $A = 0.375375375…$.

So we get: $1000A = 375 + 0.375375375… = 375 + A \Rightarrow 1000A = 375 + A$.

Thus, we get: $1000A - A = 375 \Rightarrow 999A = 375 \Rightarrow A = \dfrac{375}{999}$.

So we get: $0.375375375… = \dfrac{375}{999}$.

0.1.   0.476547654765…

We can see that the four digits 4765 are repeating.

So assuming: $A = 0.476547654765…$, we can set: $10000A = 4765.476547654765…$.

That is, we get: $10000A = 4765 + 0.476547654765…$.

So we get: $10000A = 4765 + A \Rightarrow 10000A - A = 9999A = 4765 \Rightarrow A = \dfrac{4765}{9999}$.

Thus, we get: $0.476547654765… = \dfrac{4765}{9999}$.

0.2. 2.0496496496...

Assuming first, $A = 2.0496496496...$, we can put it this way: $A = 2 + 0.0496496496....$

And assuming next, $B = 0.496496496...$, we can put A this way, too: $A = 2 + \dfrac{B}{10}$.

And in B, we can see the three digits 496 repeating.

So to begin with, we can get: $1000B = 496.496496496... = 496 + B$.

Thus, we get: $999B = 496 \Rightarrow B = \dfrac{496}{999}$.

And we have: $A = 2 + \dfrac{B}{10}$. So we get: $A = 2 + \dfrac{496}{9990}$.

Meanwhile:

$$2 + \frac{496}{9990} = \frac{2(10000 - 10)}{9990} + \frac{496}{9990} = \frac{20000 - 20 + 496}{9990}$$

$$\frac{19980 + 496}{9990} = \frac{20000 + 476}{9990} = \frac{20476}{9990}$$

Thus, we get: $A = \dfrac{20476}{9990}$.

And of course, we can put it this way, too: $A = \dfrac{10238}{4995}$.

54

0.3.   7.2060706070607...

Assuming first, **A = 7.2060706070607...**, we can put it this way:

A = 7.2 + 0.0060706070607....

And assuming next, **B = 0.0060706070607...**, we can set: **A = 7.2 + B**, and we can see four digits repeating in **B**, and they are 0607.

So assuming now, **C = 0.060706070607...**, we can set: $B = \dfrac{C}{10}$.

Then, we can get: $A = 7.2 + 0.0060706070607... = 7.2 + \dfrac{B}{10} \Rightarrow A = 7.2 + \dfrac{B}{10}$.

So beginning with **C**, we get: **10000C = 607.060706070607... = 607 + C**.

Thus, we get: $10000C - C = 9999C = 607 \Rightarrow C = \dfrac{607}{9999}$.

Next, we know: $B = \dfrac{C}{10}$. So we get: $B = \dfrac{607}{99990}$.

Next, we have: $A = 7.2 + \dfrac{B}{10}$. Thus, we get: $A = 7.2 + \dfrac{607}{999900}$.

Meanwhile:

$7.2 + \dfrac{607}{999900} = 7 + 0.2 + \dfrac{607}{999900}$.

$7 \cdot 999900 = 7(1000000 - 100) = 7 \cdot 10^6 - 700.$

$0.2 \cdot 999900 = 2 \cdot 99990 = 2(100000 - 10) = 2 \cdot 10^5 - 20.$

$-700 - 20 + 607 = -100 - 20 + 7 = -113$, and $200 - 113 = 87$.

So we get: $7 \cdot 10^6 - 700 + 2 \cdot 10^5 - 20 + 607 = 7 \cdot 10^6 + 2 \cdot 10^5 - 113$
$= 7 \cdot 10^6 + 1 \cdot 10^5 + 99800 + 200 - 113 = 7199887.$

So we get: $A = \dfrac{7199887}{999900}$.

1. **Some fractional numbers can be put in decimals. For instance:**

1/7 = 0.142857142857142857...
2/7 = 0.285714285714285714...
3/7 = 0.428571428571428571...

1.0. Find the 17th digit below the point in the decimal equivalent of 8/7.

We know: $\dfrac{8}{7} = 1 + \dfrac{1}{7}$, and we have: $\dfrac{1}{7} = 0.142857142857142857...$

So we get: $\dfrac{8}{7} = 1.142857142857142857...$

And the seventeenth digit below the dot is 5.

1.1. Find the 24th digit below the point in the decimal equivalent of 30/7.

We know: $\dfrac{30}{7} = 4 + \dfrac{2}{7}$, and we have: $\dfrac{2}{7} = 0.285714285714285714...$

So we get: $\dfrac{30}{7} = 4.\underline{285714}285714285714...$

And we can see six digits repeating, which are 285714.

And 4 divides 24, and the last digit in the six digit is 4.

So the twenty fourth digit below the dot is 4.

2.0 Convert 5/33 into the decimal equivalent, and find the thousandth digit below the point in the decimal.

Doing the direct division, we get: $\dfrac{5}{33} = 0.15151515...$

And we can see two digits repeating, which are 15.

We know 2 divides 1000, and the last of the two digits repeating is 5.
So the thousandth digit below the dot is 5.

2.1 Convert 29/222 into the decimal equivalent, and find the thousandth digit below the point in the decimal.

Doing the direct division, we get: $\dfrac{29}{222} = 0.1306306306...$

And we can see three digits repeating, which are 306.

The repeating begins though, right after the first digit below the dot.
So we are looking at the 999th digit after the first digit.

And we know 3 divides 999, and the last digit in the three digits repeating is 6.
So the thousandth digit below the dot is 6.

3.0. Find the thousandth digit below the point in the product of 0.4122412241224... and 0.9443594435944359... Use a calculator.

We can get the digit the way below:

Convert the two decimals into their fractional equivalents.
Take the product of the two fractional numbers, and then, put the product into its decimal equivalent.
And then, find the digit.

So assuming first, $A = 0.4122412241224...$, we get:

$$10000A = 4122.412241224122... = 4122 + A \Rightarrow 9999A = 4122 \Rightarrow A = \frac{4122}{9999}.$$

And assuming next, $B = 0.9443594435944359...$, we get:

$$100000B = 94435.944359443594435... = 94435 + B \Rightarrow 99999B = 94435.$$

Thus next, taking the product of A and B, and putting the product into the decimal using a calculator, we get:

$$AB = \frac{4122}{9999} \cdot \frac{94435}{99999} = \frac{4122 \cdot 94435}{9999 \cdot 99999} = \frac{389261070}{999890001} = 0.\underline{389303893038930}...$$

We can see five digits repeating, which are 38930.
And we know 5 divides 1000, and the last of the digits repeating is 0.

So the thousandth digit below the dot is 0.

**3.1.   Find the thousandth digit below the point in the product of 1.124545454...
and 4.2879269269269...   Use a calculator.**

Assuming first, $A = 1.124545454...$, and $B = 0.004545454...$, we get: $A = 1.12 + B$.

And assuming next, $C = 0.4545454...$, we can set: $B = \dfrac{C}{1000}$, and we get:

$$100C = 45.454545... = 45 + C \Rightarrow 99C = 45 \Rightarrow C = \frac{45}{99}.$$

So we get: $B = \dfrac{C}{1000} = \dfrac{45}{99000}$. And we have: $A = 1.12 + B$.

So we get: $A = 1.12 + \dfrac{45}{99000} = \dfrac{112}{100} + \dfrac{45}{99000} = \dfrac{112 \cdot 900 + 45}{99000} = \dfrac{110880 + 45}{99000} = \dfrac{110925}{99000}$.

Next, assuming $D = 4.2879269269269...$, and $E = 0.000926926...$, we get:

$D = 4.287 + E$.

And assuming next, $F = 0.926926...$, we can set: $E = \dfrac{F}{1000}$, and we get:

$1000F = 926.926926... = 926 + F \Rightarrow 999F = 926 \Rightarrow F = \dfrac{926}{999}$.

So we get: $\dfrac{F}{1000} = \dfrac{926}{999000}$. And we have: $D = 4.287 + E$.

So we get: $D = 4.287 + \dfrac{926}{999000} = \dfrac{4287}{1000} + \dfrac{926}{999000} = \dfrac{4287 \cdot 999 + 926}{999000} = \dfrac{4283639}{999000}$.

Thus next, taking the product of A and D, and putting the product into the decimal using a calculator, we get:

$AD = (110925/99000)(4283639/999000) = 475162656075/98901000000$
$= 4.80442721585221585221585...$

We can see six digits repeating, which are 215852, which begin repeating however, right after the sixth digit below the dot.
So we are looking at the 994th digit after the sixth digit.

And we have: $994 = 165 \cdot 6 + 4$.
So the digit we are after is the fourth digit in 215852.

And thus, the thousandth digit below the dot is 8.

Some rational numbers look like irrational numbers when they get converted to decimal equivalents. That's because the calculator has limits in the number of the digits that can be displayed.

For instance, $\dfrac{49356012946}{3332666700000}$ is a rational number, and has to have repeating digits, because 3 divides the denominator, but cannot divide the numerator.

Using a calculator, we can get:

$\dfrac{49356012946}{3332666700000} = 0.0148097656888401111338256537 92502$, which seems to have no

repeating digits, and thus, looks like an irrational number.

It is rational, of course. That's simply because a rational number is a ratio between two integers, and 3332666700000 and 49356012946 are integers.

4. Irrational Numbers 1

To begin with, what are irrational numbers?

The word 'irrational' in 'irrational number' is the antonym of the adjective of ratio, and has little to do with the dictionary meaning of unreasonable.
What then, is an irrational number?

We know a rational number can be called an integer ratio. More specifically, a rational number is a ratio between two integers as 2/3, 3/5, -3/7, 0.12 = 12/100 = 6/50 = 3/25, etc.

So if a number is rational, it is an integer ratio, and can be expressed by a ratio between two integers. If however, we cannot put a number in an integer ratio, it is not rational, and said to be irrational. So an irrational number cannot be expressed by a ratio between two integers, and thus, is not like 1/2, 0.3, or 3/7.

And irrational numbers can be quickly called *irrationals*.
So calling a number an irrational, we mean it's an irrational number.
And a typical example of an irrational is: $\sqrt{2}$, called a square root of 2.
What then, is $\sqrt{2}$?

If we get 2 multiplying 1 by a positive number twice, the number is $\sqrt{2}$.
So multiplying 1 by $\sqrt{2}$ twice, we get 2. That is, we have: $1 \times \sqrt{2} \times \sqrt{2} = 2$.

Why does the number multiplied twice have to be a positive number though?

That's because $\sqrt{2}$ is positive.

And of course, we can get 2 if we multiply 1 by $-\sqrt{2}$ twice, too.

So in short, multiplying 1 by $\pm\sqrt{2}$ twice, we get 2.

That is to say that we get: $(\pm\sqrt{2})^2 = 2$. And we read $\pm\sqrt{2}$ as plus or minus $\sqrt{2}$.

And we want to note that any number inside the sign $\sqrt{}$ is positive or 0, and <u>cannot be</u> <u>negative</u>. And we usually call the sign a square root sign.

So in short, what's inside the square root sign is ≥ 0, and can <u>never</u> be < 0. Why not?

That's because we get the number inside the square root sign multiplying 1 by a number twice, and if 1 is multiplied by a number twice, the product is ≥ 0, and cannot be < 0.

So *what's inside the square root sign* is ≥ 0, and can *never* be *negative*.

How come then, $\sqrt{2}$ is an irrational?

That's because it cannot be put in an integer ratio, that is, a ratio between two integers. We will get to see why it is the case doing one of the examples on irrationals. Showing though, it is an irrational, we can use the fact below:

• If <u>a number</u> is <u>rational</u>, it <u>can be</u> put in <u>a ratio between two integers prime to each other</u>.

That's because if we simplify such a ratio to its simplest, the ratio simplest has to be between two integers that are prime to each other. If <u>no divisor other than 1 is common</u> to integers, the integers are said to be <u>prime to each other</u>. So for instance, 12 and 25 are prime to each other, because 1 is the only divisor common to the two integers. That is, other than 1, there is no integer that can divide 12 and 25 both.

We have: $12/25 = 48/100 = 0.48$. So 0.48 is a rational number, and can be put in a ratio between two integers prime to each other.

And we have many kinds in irrationals. Among those, we have the circular ratio denoted by π, which is read as pi, and is 3.141592..., which is usually approximated to be 3.14.

And another example irrational can be the Euler's number, which is denoted by e, and is 2.718181828459045..., which is usually approximated to be 2.718.

And such a number as $\sqrt{2}$ is called a radical, too, specifically called a radical of degree two or a second degree radical.

And we call a *radical* a *root*, too. So $\sqrt{2}$ is called a second root or a square root.

And in fact, a radical has its degree.

Expressing a radical in general, we can put it this way: $\sqrt[n]{a}$, where n is an integer ≥ 2, and indicates the degree of the radical. So for instance, $\sqrt[3]{a}$ is a radical of degree 3, and the degree of $\sqrt{5}$ is 2, because in fact, $\sqrt{5} = \sqrt[2]{5}$. What number then, is $\sqrt[n]{a}$?

> If we get a multiplying 1 by a positive number n times, the number is: $\sqrt[n]{a}$, which is called the nth root of a or the nth radical of a.

- And the statement above is the <u>definition</u> of $\sqrt[n]{a}$.

So for instance, multiplying 1 by $\sqrt[3]{2}$ three times, we get 2. And we call it the third root of 2 or the third radical of 2.

Usually though, we call $\sqrt[3]{2}$ a *cube* root of 2, and thus, call $\sqrt[3]{}$ a cube root sign.

Unlike the case of the square root sign, we can put any real number inside the cube root sign. So what's inside the cube root sign can be negative, too.

That's because we get the number inside the cube root sign multiplying 1 by a number three times, and if 1 is multiplied by a number three times, the product can be negative as well as positive or 0. For instance, we can have $\sqrt[3]{-2}$ as well as $\sqrt[3]{2}$.

So <u>what's inside the cube root sign</u> can be <u>any real number</u>.

And covered in this book are the radicals of degrees 2 and 3. For more details on radicals and all the degrees, refer to the book called POWERS & LOGARITHMS.

So let's now move on to the arithmetic on irrationals in those two kinds.
First, the following three basic laws on arithmetic apply:

• Commutative Law

$1 + 2 = 2 + 1$, and in general, $a + b = b + a$.

$2 \times 3 = 3 \times 2$, and in general, $a \times b = b \times a$.

• Associative Law

$1 + 2 + 3 = (1 + 2) + 3 = 1 + (2 + 3)$, and in general, $a + b + c = a + b + c = a + b + c$.

$2 \times 3 \times 4 = (2 \times 3) \times 4 = 2 \times (3 \times 4)$, and in general, $a \times b \times c = (a \times b) \times c = a \times (b \times c)$.

• Distributive Law

$2(3 + 4) = 2 \times 3 + 2 \times 4$, and in general, $a(b + c) = ab + ac$

By the way, doing a multiplication, we don't usually use the operator $\times$, and instead, we just use a dot, $\cdot$, as $2 \times 3 = 2 \cdot 3$, or use nothing as $a \times b = ab$, if no ambiguity is expected.

So to begin with, adding together 1 and $\sqrt{2}$, we just get: $1 + \sqrt{2} = \sqrt{2} + 1$.
And in general, adding together a and $\sqrt{b}$, we just get: $a + \sqrt{b} = \sqrt{b} + a$.

Next, adding together $\sqrt{2}$ and $\sqrt{3}$, we just get: $\sqrt{2} + \sqrt{3}$.
And in general, adding together $\sqrt{a}$ and $\sqrt{b}$, we just get: $\sqrt{a} + \sqrt{b}$.

Next, adding together $\sqrt{2}$ and $\sqrt{2}$, we get: $2\sqrt{2}$. That is, we get: $\sqrt{2} + \sqrt{2} = 2\sqrt{2}$.
That's because two of $\sqrt{2}$ s is: 2 times $\sqrt{2}$, which is: $2\sqrt{2}$.

So in general, adding together n of $\sqrt{a}$ s, we get: $n\sqrt{a}$.

Thus next, adding together $\sqrt{2}$ and $3\sqrt{2}$, we get: $\sqrt{2} + 3\sqrt{2} = 4\sqrt{2}$.

And also, adding together $2\sqrt{3}$ and $5\sqrt{3}$, we get: $2\sqrt{3} + 5\sqrt{3} = 7\sqrt{3}$.

So in general, adding together $n\sqrt{a}$ and $m\sqrt{a}$, we just get: $n\sqrt{a} + m\sqrt{a} = (m+n)\sqrt{a}$.

Thus, for more instance, we get: $\sqrt{5} + 2\sqrt{5} = 3\sqrt{5}$, and $2\sqrt{7} + 5\sqrt{7} = 7\sqrt{7}$.

Next, adding together $\sqrt{2}$ and $2\sqrt{3}$, we just get: $\sqrt{2} + 2\sqrt{3}$.

And also, adding together $2\sqrt{2}$ and $5\sqrt{3}$, we just get: $2\sqrt{2} + 5\sqrt{3}$.

And in general, adding together $c\sqrt{a}$ and $d\sqrt{b}$, we just get: $c\sqrt{a} + d\sqrt{b}$.

Next, moving on to subtractions, and subtracting 1 from $\sqrt{3}$, we just get: $\sqrt{3} - 1$.

And in general, subtracting a from $\sqrt{b}$, we just get: $\sqrt{b} - a$.

Next, subtracting $\sqrt{3}$ from 1, we just get: $1 - \sqrt{3}$.

And in general, subtracting $\sqrt{a}$ from b, we just get: $b - \sqrt{a}$.

Next, subtracting $\sqrt{2}$ from $\sqrt{3}$, we just get: $\sqrt{3} - \sqrt{2}$.

And in general, subtracting $\sqrt{a}$ from $\sqrt{b}$, we just get: $\sqrt{b} - \sqrt{a}$.

Next, subtracting $3\sqrt{2}$ from $2\sqrt{3}$, we just get: $2\sqrt{3} - 3\sqrt{2}$.

And in general, subtracting $n\sqrt{a}$ from $m\sqrt{b}$, we just get: $m\sqrt{b} - n\sqrt{a}$.

Next, subtracting $3\sqrt{2}$ from $5\sqrt{2}$, we get: $5\sqrt{2} - 3\sqrt{2} = 2\sqrt{2}$.

And also, subtracting $5\sqrt{2}$ from $2\sqrt{2}$, we get: $2\sqrt{2} - 5\sqrt{2} = -3\sqrt{2}$.

And in general, subtracting $n\sqrt{a}$ from $m\sqrt{a}$, we just get: $m\sqrt{a} - n\sqrt{a} = (m - n)\sqrt{a}$.

Next, moving on to multiplications, and multiplying 2 by $\sqrt{3}$, we get: $2 \times \sqrt{3} = 2\sqrt{3}$.

And in general, multiplying a by $\sqrt{b}$, we get: $a\sqrt{b}$.

Next, multiplying $\sqrt{2}$ by $\sqrt{3}$, we get: $\sqrt{2}\sqrt{3} = \sqrt{2 \cdot 3} = \sqrt{6}$.

And in general, taking the product of $\sqrt{a}$ and $\sqrt{b}$, we get: $\sqrt{a}\sqrt{b} = \sqrt{ab}$.

- How can we get: $\sqrt{a}\sqrt{b} = \sqrt{ab}$, though?

By definition, we have: $1 \cdot \sqrt{a}\sqrt{a} = \sqrt{a}\sqrt{a} = a$, and $1 \cdot \sqrt{b}\sqrt{b} = \sqrt{b}\sqrt{b} = b$.

So by definition, we get: $1 \cdot \sqrt{ab}\sqrt{ab} = \sqrt{ab}\sqrt{ab} = ab$.

And by definition again, we get: $1 \cdot \sqrt{a}\sqrt{b} \cdot \sqrt{a}\sqrt{b} = \sqrt{a}\sqrt{b}\sqrt{a}\sqrt{b} = \sqrt{a}\sqrt{a}\sqrt{b}\sqrt{b} = ab$.

Thus, we get: $\sqrt{a}\sqrt{b} = \sqrt{ab}$.

So next, multiplying $2\sqrt{2}$ by $5\sqrt{3}$, we get: $2\sqrt{2} \cdot 5\sqrt{3} = 2 \cdot 5\sqrt{2}\sqrt{3} = 10\sqrt{6}$.

And in general, multiplying $c\sqrt{a}$ by $d\sqrt{b}$, we get: $c\sqrt{a} \cdot d\sqrt{b} = cd\sqrt{a}\sqrt{b} = cd\sqrt{ab}$.

Then, what if we multiply $2 + 3\sqrt{2}$ by $3 - 4\sqrt{5}$?

We can use the distributive law. Then, we get:

$$(2+3\sqrt{2})(3-4\sqrt{5}) = 2\cdot 3 - 2\cdot 4\sqrt{5} + 3\cdot 3\sqrt{2} - 3\sqrt{2}\cdot 4\sqrt{5} = 6 - 8\sqrt{5} + 9\sqrt{2} - 12\sqrt{10}.$$

And by the same token, we can get:

$$(\sqrt{3}+3\sqrt{2})(2\sqrt{7}-4\sqrt{5}) = 2\sqrt{3}\sqrt{7} - 4\sqrt{3}\sqrt{5} + 6\sqrt{2}\sqrt{7} - 12\sqrt{2}\sqrt{5} = 2\sqrt{21} - 4\sqrt{15} + 6\sqrt{14}.$$

And in general, we have:

$$(a\sqrt{x}+b\sqrt{y})(s\sqrt{u}+t\sqrt{v}) = as\sqrt{ux} + at\sqrt{vx} + bs\sqrt{uy} + bt\sqrt{vy}.$$

Examples 1 in Irrationals

Simplify each radical below or put each in a mixed radical if it can be done so. A mixed radical is a radical multiplied by a rational such as $2\sqrt{3}$ and $4\sqrt[3]{2}$.

0. $\sqrt{4}$ 1. $\sqrt{8}$ 2. $\sqrt{-2}$ 3. $\sqrt{12}$ 4. $\sqrt[3]{8}$

5. $\sqrt[3]{32}$ 6. $\sqrt[3]{-8}$ 7. $\sqrt{1.69}$ 8. $\sqrt[3]{5.6}$ 9. $\sqrt[3]{0.016}$

A. $\sqrt[3]{-3.2}$ B. $\sqrt[3]{270}$ C. $\sqrt[3]{0.081}$ D. $\sqrt[3]{-16}$

Suggestions or Solutions
To the Examples 1 in Irrationals

0. $\sqrt{4} = \sqrt{2^2} = 2$, because $1 \times \sqrt{4} \times \sqrt{4} = 4$, which is what's inside the square root sign, and we have: $2^2 = 4$, too, which means we have: $\mathbf{1 \cdot 2 \cdot 2 = 4}$. So we get: $\sqrt{4} = 2$. What then, about: $\mathbf{1 \cdot (-2) \cdot (-2) = 4}$?

We know $\sqrt{4}$ is positive. So we don't get this: $\sqrt{4} = \mathbf{-2}$.

And by the same token, we get: $\sqrt{9} = \sqrt{3^2} = 3$, $\sqrt{16} = \sqrt{4^2} = 4$, $\sqrt{25} = \sqrt{5^2} = 5$, etc.

1. $\sqrt{8} = \sqrt{4 \cdot 2} = \sqrt{4}\sqrt{2} = 2\sqrt{2}$. How come though, do we get: $\sqrt{8} = \sqrt{4 \cdot 2} = \sqrt{4}\sqrt{2}$?

By definition, we get: $\mathbf{1 \times \sqrt{8} \times \sqrt{8} = 8}$.
Andy by definition again, we get: $\mathbf{1 \times \sqrt{4 \cdot 2} \times \sqrt{4 \cdot 2} = 4 \cdot 2 = 8}$.
And also, we get: $\mathbf{1 \times \sqrt{4}\sqrt{2} \times \sqrt{4}\sqrt{2} = \sqrt{4} \times \sqrt{4} \times \sqrt{2} \times \sqrt{2} = 4 \times 2 = 8}$.
So we get: $\sqrt{8} = \sqrt{4 \cdot 2} = \sqrt{4}\sqrt{2}$.

And by the same token, we get:

$$\sqrt{21} = \sqrt{3 \cdot 7} = \sqrt{3}\sqrt{7} \,, \quad \sqrt{15} = \sqrt{3 \cdot 5} = \sqrt{5}\sqrt{3} \,, \quad \sqrt{54} = \sqrt{6 \cdot 9} = \sqrt{9}\sqrt{6} = 3\sqrt{6}$$

$$\sqrt{21} = \sqrt{3 \cdot 7} = \sqrt{3}\sqrt{7} \,, \quad \sqrt{24} = \sqrt{6 \cdot 4} = \sqrt{6}\sqrt{4} = 2\sqrt{6} \,, \quad \sqrt{72} = \sqrt{36 \cdot 2} = \sqrt{36}\sqrt{2} = 6\sqrt{2}$$

2. $\sqrt{-2}$ cannot be made within the set of all real numbers, and thus, is not allowed, because no real number can be squared to be a negative number.

3. $\sqrt{12} = \sqrt{4 \cdot 3} = \sqrt{4}\sqrt{3} = 2\sqrt{3}$ And by the same token, we get:

$\sqrt{18} = \sqrt{9 \cdot 2} = \sqrt{9}\sqrt{2} = 3\sqrt{2}$, $\sqrt{50} = \sqrt{25 \cdot 2} = 5\sqrt{2}$, $\sqrt{48} = \sqrt{16 \cdot 3} = 4\sqrt{3}$

$\sqrt{242} = \sqrt{121 \cdot 2} = \sqrt{11^2}\sqrt{2} = 11\sqrt{2}$, $\sqrt{432} = \sqrt{144 \cdot 3} = \sqrt{12^2}\sqrt{3} = 12\sqrt{3}$, etc.

4. $\sqrt[3]{8} = \sqrt[3]{2^3} = 2$, because $1 \times \sqrt[3]{8} \times \sqrt[3]{8} \times \sqrt[3]{8} = 8$, which is what's inside the cube root sign, and we have: $2^3 = 8$, too, which means we have: $1 \cdot 2 \cdot 2 \cdot 2 = 8$. So we get: $\sqrt[3]{8} = 2$.

And by the same token, we get:

$\sqrt[3]{27} = \sqrt[3]{3^3} = 3$, $\sqrt[3]{64} = \sqrt[3]{4^3} = 4$, $\sqrt[3]{125} = \sqrt[3]{5^3} = 5$, $\sqrt[3]{216} = \sqrt[3]{6^3} = 6$, etc.

5. $\sqrt[3]{32} = \sqrt[3]{8 \cdot 4} = \sqrt[3]{8} \cdot \sqrt[3]{4} = 2\sqrt[3]{4}$ And by the same token, we get:

$\sqrt[3]{40} = \sqrt[3]{5 \cdot 8} = \sqrt[3]{5} \cdot \sqrt[3]{8} = 2\sqrt[3]{5}$, $\sqrt[3]{54} = \sqrt[3]{27 \cdot 2} = \sqrt[3]{27} \cdot \sqrt[3]{2} = 3\sqrt[3]{2}$

$\sqrt[3]{72} = \sqrt[3]{9 \cdot 8} = \sqrt[3]{8} \cdot \sqrt[3]{9} = 2\sqrt[3]{9}$, $\sqrt[3]{81} = \sqrt[3]{27 \cdot 3} = \sqrt[3]{27} \cdot \sqrt[3]{3} = 3\sqrt[3]{3}$, $\sqrt[3]{162} = \sqrt[3]{27 \cdot 6} = 3\sqrt[3]{6}$

6. $\sqrt[3]{-8} = \sqrt[3]{(-2)^3} = -2$, because $1 \times \sqrt[3]{-8} \times \sqrt[3]{-8} \times \sqrt[3]{-8} = -8$, which is what's inside the cube root sign, and we have: $(-2)^3 = -8$, too, which means we have: $1 \cdot (-2) \cdot (-2) \cdot (-2) = -8$.

So we get: $\sqrt[3]{-8} = -2$.   And by the same token, we get:

$\sqrt[3]{-27} = \sqrt[3]{(-3)^3} = -3$, $\sqrt[3]{-64} = \sqrt[3]{(-4)^3} = -4$, $\sqrt[3]{-0.008} = \sqrt[3]{(-0.2)^3} = -0.2$, etc.

7. $\sqrt{1.69} = \sqrt{(1.3)^2} = 1.3$

8. $\sqrt[3]{5.6} = \sqrt[3]{8 \cdot 0.7} = \sqrt[3]{8} \cdot \sqrt[3]{0.7} = 2\sqrt[3]{0.7}$

9. $\sqrt[3]{0.016} = \sqrt[3]{8 \cdot 0.002} = \sqrt[3]{8} \cdot \sqrt[3]{0.001} \cdot \sqrt[3]{2} = 2 \cdot 0.1\sqrt[3]{2} = 0.2\sqrt[3]{2}$

A. $\sqrt[3]{-3.2} = \sqrt[3]{-8 \cdot 0.4} = \sqrt[3]{-8} \cdot \sqrt[3]{0.4} = -2\sqrt[3]{0.4}$

B. $\sqrt[3]{270} = \sqrt[3]{27 \cdot 10} = \sqrt[3]{27}\sqrt[3]{10} = 3\sqrt[3]{10}$

C. $\sqrt[3]{0.081} = \sqrt[3]{0.001 \cdot 81} = \sqrt[3]{0.001}\sqrt[3]{81} = 0.1\sqrt[3]{27 \cdot 3} = 0.1\sqrt[3]{3^3 3} = 0.3\sqrt[3]{3}$

D. $\sqrt[3]{-160} = \sqrt[3]{-8 \cdot 20} = \sqrt[3]{-2^3 20} = -2\sqrt[3]{20}$

Note that: $-2^3 = (-2)^3 = -8$, but $-2^2 \neq (-2)^2$, because $-2^2 = -4$, and $(-2)^2 = 4$.

Examples 2 in Irrationals

0. $\sqrt{4} + \sqrt{9}$ 1. $\sqrt{4} + \sqrt{8}$ 2. $\sqrt{18} + \sqrt{8}$

3. $\sqrt{45} + \sqrt{20}$ 4. $\sqrt{24} + \sqrt{54}$ 5. $\sqrt{21} + \sqrt{35}$

6. $\sqrt{150} - \sqrt{24}$ 7. $\sqrt{72} - \sqrt{18}$ 8. $\sqrt{27} - \sqrt{12}$

9. $\sqrt{18} - \sqrt{162}$ A. $\sqrt{625} - \sqrt{125}$ B. $\sqrt{48} - \sqrt{108}$

C. $\sqrt{242} - \sqrt{338}$ D. $\sqrt{75} - \sqrt{432}$ E. $\sqrt[3]{27} - \sqrt[3]{64}$

F. $\sqrt[3]{250} - \sqrt[3]{432}$ G. $\sqrt[3]{108} - \sqrt[3]{32}$ H. $\sqrt[3]{72} - \sqrt[3]{243}$

I. $\sqrt[3]{-27} - \sqrt[3]{-64}$ J. $\sqrt[3]{-0.008} - \sqrt{1.69}$ K. $\sqrt[3]{1.6} - \sqrt[3]{0.0016}$

L. $\sqrt[3]{270} - \sqrt[3]{-80}$ M. $\sqrt[3]{0.016} - \sqrt[3]{16}$

Suggestions or Solutions
To the Examples 2 in Irrationals

0. $\sqrt{4}+\sqrt{9}=2+3=5$

1. $\sqrt{4}+\sqrt{8}=2+2\sqrt{2}$

2. $\sqrt{18}+\sqrt{8}=\sqrt{9\cdot 2}+2\sqrt{2}=3\sqrt{2}+2\sqrt{2}=5\sqrt{2}$

3. $\sqrt{45}+\sqrt{20}=\sqrt{9\cdot 5}+\sqrt{4\cdot 5}=3\sqrt{5}+2\sqrt{5}=5\sqrt{5}$

4. $\sqrt{24}+\sqrt{54}=\sqrt{4\cdot 6}+\sqrt{6\cdot 9}=2\sqrt{6}+3\sqrt{6}=5\sqrt{6}$

5. $\sqrt{21}+\sqrt{35}=\sqrt{3\cdot 7}+\sqrt{5\cdot 7}=\sqrt{3}\sqrt{7}+\sqrt{5}\sqrt{7}=\sqrt{7}(\sqrt{3}+\sqrt{5})$

Note however, it's just an example where you can see how we can change an expression (without changing its value, of course), so you can just leave $\sqrt{21}+\sqrt{35}$ as is.

6. $\sqrt{150}-\sqrt{24}=\sqrt{25\cdot 6}-\sqrt{6\cdot 4}=5\sqrt{6}-2\sqrt{6}=3\sqrt{6}$

7. $\sqrt{72}-\sqrt{18}=\sqrt{36\cdot 2}-\sqrt{9\cdot 2}=6\sqrt{2}-3\sqrt{2}=3\sqrt{2}$ We can put it the way below, too:

$\sqrt{72}-\sqrt{18}=\sqrt{18\cdot 4}-\sqrt{18}=2\sqrt{18}-\sqrt{18}=\sqrt{18}=\sqrt{9\cdot 2}=3\sqrt{2}$

8. $\sqrt{27}-\sqrt{12}=\sqrt{9\cdot 3}-\sqrt{4\cdot 3}=3\sqrt{3}-2\sqrt{3}=\sqrt{3}$

9. $\sqrt{18} - \sqrt{162} = \sqrt{9 \cdot 2} - \sqrt{81 \cdot 2} = 3\sqrt{2} - 9\sqrt{2} = -6\sqrt{2}$,

A. $\sqrt{625} - \sqrt{125} = \sqrt{25^2} - \sqrt{25 \cdot 5} = 25 - 5\sqrt{5} = 5(5 - \sqrt{5}) = 5(\sqrt{5}\sqrt{5} - \sqrt{5})$

$= 5\sqrt{5}(\sqrt{5} - 1) = 5\sqrt{5}(\sqrt{5} - \frac{\sqrt{5}}{\sqrt{5}}) = 5\sqrt{5}\sqrt{5}(1 - \frac{1}{\sqrt{5}}) = 25(1 - \frac{1}{\sqrt{5}}) \dots$

Usually however, we just leave it as $25 - 5\sqrt{5}$.

B. $\sqrt{48} - \sqrt{108} = \sqrt{16 \cdot 3} - \sqrt{36 \cdot 3} = 4\sqrt{3} - 6\sqrt{3} = -2\sqrt{3}$

C. $\sqrt{242} - \sqrt{338} = \sqrt{121 \cdot 2} - \sqrt{169 \cdot 2} = \sqrt{11^2}\sqrt{2} - \sqrt{13^2}\sqrt{2} = 11\sqrt{2} - 13\sqrt{2} = -2\sqrt{2}$

D. $\sqrt{75} - \sqrt{432} = \sqrt{25 \cdot 3} - \sqrt{144 \cdot 3} = 5\sqrt{3} - \sqrt{12^2}\sqrt{3} = -7\sqrt{3}$

E. $\sqrt[3]{27} - \sqrt[3]{64} = \sqrt[3]{3^3} - \sqrt[3]{4^3} = 3 - 4 = -1$

F. $\sqrt[3]{250} - \sqrt[3]{432} = \sqrt[3]{125 \cdot 2} - \sqrt[3]{216 \cdot 2} = \sqrt[3]{5^3 2} - \sqrt[3]{6^3 2} = 5\sqrt[3]{2} - 6\sqrt[3]{2} = -\sqrt[3]{2}$

G. $\sqrt[3]{108} - \sqrt[3]{32} = \sqrt[3]{27 \cdot 4} - \sqrt[3]{8 \cdot 4} = 3\sqrt[3]{4} - 2\sqrt[3]{4} = \sqrt[3]{4}$

H. $\sqrt[3]{72} - \sqrt[3]{243} = \sqrt[3]{9 \cdot 8} - \sqrt[3]{27 \cdot 9} = 2\sqrt[3]{9} - 3\sqrt[3]{9} = -\sqrt[3]{9}$

I. $\sqrt[3]{-27} - \sqrt[3]{-64} = \sqrt[3]{(-3)^3} - \sqrt[3]{(-4)^3} = -3 - (-4) = 1$

J. $\sqrt[3]{-0.008} - \sqrt{1.69} = \sqrt[3]{(-0.2)^3} - \sqrt{(1.3)^2} = -0.2 - 1.3 = -1.5$

K. $\sqrt[3]{1.6} - \sqrt[3]{0.0016} = \sqrt[3]{8 \cdot 0.2} - \sqrt[3]{8 \cdot 0.0002} = 2\sqrt[3]{0.2} - \sqrt[3]{0.008} \cdot \sqrt[3]{0.2}$

$= 2\sqrt[3]{0.2} - 0.2\sqrt[3]{0.2} = 1.8\sqrt[3]{0.2}$

L. $\sqrt[3]{270} - \sqrt[3]{-80} = \sqrt[3]{270} + \sqrt[3]{80} = \sqrt[3]{27 \cdot 10} + \sqrt[3]{8 \cdot 10} = 3\sqrt[3]{10} + 2\sqrt[3]{10} = 5\sqrt[3]{10}$

M. $\sqrt[3]{0.016} - \sqrt[3]{16} = \sqrt[3]{0.001 \cdot 16} - \sqrt[3]{16} = \sqrt[3]{0.001}\sqrt[3]{16} - \sqrt[3]{16} = 0.1\sqrt[3]{16} - \sqrt[3]{16}$

$= (0.1 - 1)\sqrt[3]{16} = -0.9\sqrt[3]{8 \cdot 2} = -1.8\sqrt[3]{2}$

Examples 3 in Irrationals

0. $\sqrt{3} \times \sqrt{7}$

1. $\sqrt[3]{3} \times \sqrt[3]{4}$

2. $\sqrt{4}\sqrt{8}$

3. $\sqrt{18}\sqrt{8}$

4. $\sqrt{45}\sqrt{20}$

5. $\sqrt{24}\sqrt{14}$

6. $\sqrt{21}\sqrt{45}$

7. $\sqrt{150}\sqrt{28}$

8. $\sqrt{72}\sqrt{36}$

9. $\sqrt{27}\sqrt{24}$

A. $\sqrt{54}\sqrt{162}$

B. $\sqrt{625}\sqrt{125}$

C. $\sqrt{48}\sqrt{216}$

D. $\sqrt{363}\sqrt{338}$

E. $\sqrt{150}\sqrt{432}$

F. $\sqrt[3]{54}\sqrt[3]{192}$

G. $\sqrt[3]{250}\sqrt[3]{432}$

H. $\sqrt[3]{108}\sqrt[3]{32}$

I. $\sqrt[3]{72}\sqrt[3]{243}$

J. $\sqrt[3]{-54}\sqrt[3]{-128}$

K. $\sqrt[3]{-0.008} \cdot \sqrt[3]{125}$

L. $\sqrt[3]{1.6} \cdot \sqrt[3]{16}$

M. $\sqrt[3]{270}\sqrt[3]{-80}$

N. $\sqrt[3]{0.016} \cdot \sqrt[3]{16}$

Suggestions or Solutions
To the Examples 3 in Irrationals

0. $\sqrt{3} \times \sqrt{7} = \sqrt{3} \cdot \sqrt{7} = \sqrt{3}\sqrt{7} = \sqrt{3 \times 7} = \sqrt{3 \cdot 7} = \sqrt{21}$

1. $\sqrt[3]{3} \times \sqrt[3]{4} = \sqrt[3]{3} \cdot \sqrt[3]{4} = \sqrt[3]{3}\sqrt[3]{4} = \sqrt[3]{3 \times 4} = \sqrt[3]{3 \cdot 4} = \sqrt[3]{12}$

2. $\sqrt{4}\sqrt{8} = 2 \cdot 2\sqrt{2} = 4\sqrt{2}$, or $\sqrt{4}\sqrt{8} = \sqrt{32} = \sqrt{16 \cdot 2} = 4\sqrt{2}$

3. $\sqrt{18}\sqrt{8} = 3\sqrt{2} \cdot 2\sqrt{2} = 6 \cdot 2 = 12$, or $\sqrt{18}\sqrt{8} = \sqrt{144} = \sqrt{12^2} = 12$

4. $\sqrt{45}\sqrt{20} = 3\sqrt{5} \cdot 2\sqrt{5} = 6 \cdot 5 = 30$, or $\sqrt{45}\sqrt{20} = \sqrt{900} = \sqrt{30^2} = 30$

5. $\sqrt{24}\sqrt{14} = 2\sqrt{6} \cdot \sqrt{14} = 2\sqrt{6 \cdot 14} = 2\sqrt{2 \cdot 3 \cdot 2 \cdot 7} = 2\sqrt{4 \cdot 3 \cdot 7} = 4\sqrt{3 \cdot 7} = 4\sqrt{21}$

6. $\sqrt{21}\sqrt{45} = \sqrt{3 \cdot 7} \cdot \sqrt{5 \cdot 9} = 3\sqrt{3 \cdot 5 \cdot 7} = 3\sqrt{105}$

7. $\sqrt{150}\sqrt{28} = \sqrt{25 \cdot 6}\sqrt{7 \cdot 4} = 5\sqrt{6} \cdot 2\sqrt{7} = 10\sqrt{6 \cdot 7} = 10\sqrt{42}$

8. $\sqrt{72}\sqrt{36} = \sqrt{36 \cdot 2}\sqrt{36} = 36\sqrt{2}$

9. $\sqrt{27}\sqrt{24} = \sqrt{9 \cdot 3}\sqrt{4 \cdot 6} = 3\sqrt{3} \cdot 2\sqrt{3 \cdot 2} = 3 \cdot 2 \cdot \sqrt{3}\sqrt{3 \cdot 2} = 6 \cdot 3\sqrt{2} = 18\sqrt{2}$

A. $\sqrt{54}\sqrt{162} = \sqrt{9\cdot6}\sqrt{81\cdot2} = 3\sqrt{6}\cdot9\sqrt{2} = 18\sqrt{12} = 18\sqrt{4\cdot3} = 36\sqrt{3}$,

B. $\sqrt{625}\sqrt{125} = \sqrt{25^2}\sqrt{25\cdot5} = 25\cdot5\sqrt{5} = 125\sqrt{5}$

In fact, $625 = 125\cdot5$, so $\sqrt{625}\sqrt{125} = \sqrt{125\cdot5}\sqrt{125} = 125\sqrt{5}$

C. $\sqrt{48}\sqrt{216} = \sqrt{16\cdot3}\sqrt{36\cdot6} = 4\sqrt{3}\cdot6\sqrt{6} = 24\cdot3\sqrt{2} = 72\sqrt{2}$

D. $\sqrt{363}\sqrt{338} = \sqrt{121\cdot3}\sqrt{169\cdot2} = \sqrt{11^2}\sqrt{3}\sqrt{13^2}\sqrt{2} = 11\cdot13\sqrt{6} = 143\sqrt{6}$

E. $\sqrt{150}\sqrt{432} = \sqrt{50\cdot3}\sqrt{144\cdot3} = \sqrt{25\cdot6}\sqrt{12^2}\sqrt{3} = 5\sqrt{6}\cdot12\sqrt{3} = 60\cdot3\sqrt{2} = 180\sqrt{2}$

F. $\sqrt[3]{54}\sqrt[3]{192} = \sqrt[3]{27\cdot2}\sqrt[3]{64\cdot3} = 3\sqrt[3]{2}\cdot4\sqrt[3]{3} = 12\sqrt[3]{2\cdot3} = 12\sqrt[3]{6}$

G. $\sqrt[3]{250}\sqrt[3]{432} = \sqrt[3]{125\cdot2}\sqrt[3]{216\cdot2} = \sqrt[3]{5^3 2}\sqrt[3]{6^3 2} = 5\sqrt[3]{2}\cdot6\sqrt[3]{2} = 30\sqrt[3]{4}$

H. $\sqrt[3]{108}\sqrt[3]{32} = \sqrt[3]{27\cdot4}\sqrt[3]{8\cdot4} = 3\sqrt[3]{4}\cdot2\sqrt[3]{4} = 8\sqrt[3]{16} = 8\sqrt[3]{8\cdot2} = 16\sqrt[3]{2}$

I. $\sqrt[3]{72}\sqrt[3]{243} = \sqrt[3]{9\cdot8}\sqrt[3]{27\cdot9} = 2\sqrt[3]{9}\cdot3\sqrt[3]{9} = 6\sqrt[3]{81} = 6\sqrt[3]{27\cdot3} = 18\sqrt[3]{3}$

J. $\sqrt[3]{-54}\sqrt[3]{-128} = \sqrt[3]{-27\cdot2}\sqrt[3]{-64\cdot2} = -3\sqrt[3]{2}(-4\sqrt[3]{2}) = 12\sqrt[3]{4}$

K. $\sqrt[3]{-0.008} \cdot \sqrt[3]{125} = \sqrt[3]{(-0.2)^3} \sqrt{5^3} = -0.2 \cdot 5 = -1$

L. $\sqrt[3]{1.6} \cdot \sqrt[3]{16} = \sqrt[3]{8 \cdot 0.2} \cdot \sqrt[3]{8 \cdot 2} = 2\sqrt[3]{0.2} \cdot 2\sqrt[3]{2} = 4\sqrt[3]{0.4}$

M. $\sqrt[3]{270}\sqrt[3]{-80} = \sqrt[3]{27 \cdot 10} \cdot \sqrt[3]{-8 \cdot 10} = 3\sqrt[3]{10}(-2\sqrt[3]{10}) = -6\sqrt[3]{100}$

N. $\sqrt[3]{0.016} \cdot \sqrt[3]{16} = \sqrt[3]{0.001 \cdot 16} \cdot \sqrt[3]{16} = 0.1\sqrt[3]{16} \cdot \sqrt[3]{16} = 0.1\sqrt[3]{16 \cdot 16}$

$= 0.1\sqrt[3]{64 \cdot 4} = 0.1\sqrt[3]{64}\sqrt[3]{4} = 0.1 \cdot 4\sqrt[3]{4} = 0.4\sqrt[3]{4}$

5. Irrational Numbers 2

Let's first, go over some of the material in the previous section.

To begin with, what are irrational numbers?

If a number is rational, it is an integer ratio, and can be expressed by a ratio between two integers. If however, we cannot put a number in an integer ratio, it is not rational, and said to be irrational. So an irrational number cannot be expressed by a ratio between two integers, and thus, is not like 1/2, 0.3, or 3/7.

And irrational numbers can be quickly called *irrationals*.
So calling a number an irrational, we mean it's an irrational number.
And a typical example of an irrational is: $\sqrt{2}$, called a square root of 2.

And we want to note that any number inside the sign $\sqrt{}$ is positive or 0, and <u>cannot be negative</u>. And we usually call the sign a square root sign.
So in short, what's inside the square root sign is ≥ 0, and can <u>never</u> be < 0.

That's because we get the number inside the square root sign multiplying 1 by a number twice, and if 1 is multiplied by a number twice, the product is ≥ 0, and cannot be < 0.

So *what's inside the square root sign* is ≥ 0, and can *never* be *negative*.

And such a number as $\sqrt{2}$ is called a radical, too, specifically called a radical of degree two or a second degree radical.
And we call a *radical* a *root*, too. So $\sqrt{2}$ is called a second root or a square root.
And in fact, a radical has its degree.

Expressing a radical in general, we can put it this way: $\sqrt[n]{a}$, where n is an integer ≥ 2, and indicates the degree of the radical. So for instance, $\sqrt[3]{a}$ is a radical of degree 3, and the degree of $\sqrt{5}$ is 2, because in fact, $\sqrt{5} = \sqrt[2]{5}$. What number then, is $\sqrt[n]{a}$?

If we get a multiplying 1 by a positive number n times, the number is: $\sqrt[n]{a}$, which is called the nth root of a or the nth radical of a.

• And the statement above is the <u>definition</u> of $\sqrt[n]{a}$.

So for instance, multiplying 1 by $\sqrt[3]{2}$ three times, we get 2. And we call it the third root of 2 or the third radical of 2.

Usually though, we call $\sqrt[3]{2}$ a *cube* root of 2, and thus, call $\sqrt[3]{}$ a cube root sign.

Unlike the case of the square root sign, we can put any real number inside the cube root sign. So what's inside the cube root sign can be negative, too.

That's because we get the number inside the cube root sign multiplying 1 by a number three times, and if 1 is multiplied by a number three times, the product can be negative as well as positive or 0. For instance, we can have $\sqrt[3]{-2}$ as well as $\sqrt[3]{2}$.
So <u>what's inside the cube root sign</u> can be <u>any real number</u>.

And covered in this book are the radicals of degrees 2 and 3.
So let's now move on to the arithmetic on irrationals in those two kinds.
First, the following three basic laws on arithmetic apply:

• Commutative Law

$1 + 2 = 2 + 1$, and in general, $a + b = b + a$.

$2 \times 3 = 3 \times 2$, and in general, $a \times b = b \times a$.

• Associative Law

$1 + 2 + 3 = (1 + 2) + 3 = 1 + (2 + 3)$, and in general, $a + b + c = a + b + c = a + b + c$.

$2 \times 3 \times 4 = (2 \times 3) \times 4 = 2 \times (3 \times 4)$, and in general, $a \times b \times c = (a \times b) \times c = a \times (b \times c)$.

• Distributive Law

$2(3 + 4) = 2 \times 3 + 2 \times 4$, and in general, $a(b + c) = ab + ac$

By the way, doing a multiplication, we don't usually use the operator x, and instead, we just use a dot, $\cdot$, as $2 \times 3 = 2 \cdot 3$, or use nothing as $a \times b = ab$, if no ambiguity is expected.

So to begin with, adding together $\sqrt{2}$ and $3\sqrt{2}$, we get: $\sqrt{2} + 3\sqrt{2} = 4\sqrt{2}$.

And also, adding together $2\sqrt{3}$ and $5\sqrt{3}$, we get: $2\sqrt{3} + 5\sqrt{3} = 7\sqrt{3}$.

So in general, adding together $n\sqrt{a}$ and $m\sqrt{a}$, we just get: $n\sqrt{a} + m\sqrt{a} = (m+n)\sqrt{a}$.

Next, adding together $\sqrt{2}$ and $2\sqrt{3}$, we just get: $\sqrt{2} + 2\sqrt{3}$.

And also, adding together $2\sqrt{2}$ and $5\sqrt{3}$, we just get: $2\sqrt{2} + 5\sqrt{3}$.

And in general, adding together $c\sqrt{a}$ and $d\sqrt{b}$, we just get: $c\sqrt{a} + d\sqrt{b}$.

Next, subtracting $3\sqrt{2}$ from $2\sqrt{3}$, we just get: $2\sqrt{3} - 3\sqrt{2}$.

And in general, subtracting $n\sqrt{a}$ from $m\sqrt{b}$, we just get: $m\sqrt{b} - n\sqrt{a}$.

Next, subtracting $3\sqrt{2}$ from $5\sqrt{2}$, we get: $5\sqrt{2} - 3\sqrt{2} = 2\sqrt{2}$.

And also, subtracting $5\sqrt{2}$ from $2\sqrt{2}$, we get: $2\sqrt{2} - 5\sqrt{2} = -3\sqrt{2}$.

And in general, subtracting $n\sqrt{a}$ from $m\sqrt{a}$, we just get: $m\sqrt{a} - n\sqrt{a} = (m-n)\sqrt{a}$.

Next, moving on to multiplications, and multiplying 2 by $\sqrt{3}$, we get: $2 \times \sqrt{3} = 2\sqrt{3}$.

And in general, multiplying a by $\sqrt{b}$, we get: $a\sqrt{b}$.

Next, multiplying $3\sqrt{5}$ by $2\sqrt{3}$, we get: $3\sqrt{5} \cdot 2\sqrt{3} = 3 \cdot 2\sqrt{5 \cdot 3} = 6\sqrt{15}$.

And in general, multiplying $c\sqrt{a}$ by $d\sqrt{b}$, we get: $c\sqrt{a} \cdot d\sqrt{b} = cd\sqrt{a}\sqrt{b} = cd\sqrt{ab}$.

Then, what if we multiply $2 + 3\sqrt{2}$ by $3 - 4\sqrt{5}$?

We can use the distributive law. Then, we get:

$$(2+3\sqrt{2})(3-4\sqrt{5})=2\cdot3-2\cdot4\sqrt{5}+3\cdot3\sqrt{2}-3\sqrt{2}\cdot4\sqrt{5}=6-8\sqrt{5}+9\sqrt{2}-12\sqrt{10}.$$

And by the same token, we can get:

$$(\sqrt{3}+3\sqrt{2})(2\sqrt{7}-4\sqrt{5})=2\sqrt{3}\sqrt{7}-4\sqrt{3}\sqrt{5}+6\sqrt{2}\sqrt{7}-12\sqrt{2}\sqrt{5}=2\sqrt{21}-4\sqrt{15}+6\sqrt{14}.$$

And in general, we have:

$$(a\sqrt{x}+b\sqrt{y})(s\sqrt{u}+t\sqrt{v})=as\sqrt{ux}+at\sqrt{vx}+bs\sqrt{uy}+bt\sqrt{vy}.$$

Next, moving on to divisions, and dividing $\sqrt{2}$ by $\sqrt{3}$, we get: $\sqrt{2}\div\sqrt{3}=\dfrac{\sqrt{2}}{\sqrt{3}}=\sqrt{\dfrac{2}{3}}.$

And in general, dividing $\sqrt{a}$ by $\sqrt{b}$, we get: $\sqrt{a}\div\sqrt{b}=\dfrac{\sqrt{a}}{\sqrt{b}}=\sqrt{\dfrac{a}{b}}.$

How come we get: $\dfrac{\sqrt{a}}{\sqrt{b}}=\sqrt{\dfrac{a}{b}}$ though?

By definition, we have: $1\cdot\sqrt{a}\sqrt{a}=\sqrt{a}\sqrt{a}=a$, and $1\cdot\sqrt{b}\sqrt{b}=\sqrt{b}\sqrt{b}=b$.

So by definition, we get: $1\cdot\sqrt{\dfrac{a}{b}}\cdot\sqrt{\dfrac{a}{b}}=\sqrt{\dfrac{a}{b}}\sqrt{\dfrac{a}{b}}=\dfrac{a}{b}.$

And by definition again, we get: $1\cdot\dfrac{\sqrt{a}}{\sqrt{b}}\cdot\dfrac{\sqrt{a}}{\sqrt{b}}=\dfrac{\sqrt{a}\sqrt{a}}{\sqrt{b}\sqrt{b}}=\dfrac{a}{b}.$

Thus, we get: $\dfrac{\sqrt{a}}{\sqrt{b}}=\sqrt{\dfrac{a}{b}}.$

So next, dividing $2\sqrt{2}$ by $5\sqrt{3}$, we get: $2\sqrt{2}\div5\sqrt{3}=\dfrac{2\sqrt{2}}{5\sqrt{3}}=\dfrac{2}{5}\cdot\dfrac{\sqrt{2}}{\sqrt{3}}=\dfrac{2}{5}\sqrt{\dfrac{2}{3}}.$

And in general, dividing $c\sqrt{a}$ by $d\sqrt{b}$, we get: $c\sqrt{a}\div d\sqrt{b}=\dfrac{c\sqrt{a}}{d\sqrt{b}}=\dfrac{c}{d}\cdot\dfrac{\sqrt{a}}{\sqrt{b}}=\dfrac{c}{d}\sqrt{\dfrac{a}{b}}.$

Next, dividing $2 + 3\sqrt{2}$ by $3 - 4\sqrt{5}$, we just get: $(2 + 3\sqrt{2}) \div (3 - 4\sqrt{5}) = \dfrac{2 + 3\sqrt{2}}{3 - 4\sqrt{5}}$.

And by the same token, we just get: $(\sqrt{3} + 3\sqrt{2}) \div (2\sqrt{7} - 4\sqrt{5}) = \dfrac{\sqrt{3} + 3\sqrt{2}}{2\sqrt{7} - 4\sqrt{5}}$.

And in general, we have: $(a\sqrt{x} + b\sqrt{y}) \div (s\sqrt{u} + t\sqrt{v}) = \dfrac{a\sqrt{x} + b\sqrt{y}}{s\sqrt{u} + t\sqrt{v}}$.

However, in all the cases above, it is *not* the case where we can use all real numbers as the letters used in all the operations above. Why not though?

Though it sounds quite natural, we need to keep in mind that *what's inside the square root sign* is positive or 0, simply because no number can be squared to be negative, and also, keep in mind that a denominator cannot be 0, since *no division by* 0 is allowed. And if in a fraction, the denominator is an irrational number, we often convert the fraction to another fraction where the denominator is a rational, because it makes more sense. And if doing so, we say that we rationalize the denominator.

So assuming now, for instance, $k = \dfrac{1}{\sqrt{2}}$, and rationalizing the denominator of k, we get:

$k = \dfrac{1}{\sqrt{2}} = \dfrac{1 \cdot \sqrt{2}}{\sqrt{2}\sqrt{2}} = \dfrac{\sqrt{2}}{2}$, which has now, a rational number as its denominator.

And rationalizing the denominator of $\dfrac{a}{\sqrt{b}}$, we can get: $\dfrac{a}{\sqrt{b}} = \dfrac{a\sqrt{b}}{\sqrt{b}\sqrt{b}} = \dfrac{a\sqrt{b}}{b} = \dfrac{a}{b}\sqrt{b}$.

What then, about this: $\dfrac{\sqrt{3}}{\sqrt{2} + 3}$?

Rationalizing the denominator of a fraction as the one above, we can use tools called *factorization identities*. And in this case, we can use this identity: $x^2 - y^2 = (x - y)(x + y)$.

Using thus, the identity above, we can get:

$$\frac{\sqrt{3}}{\sqrt{2}+3} = \frac{\sqrt{3}(\sqrt{2}-3)}{(\sqrt{2}+3)(\sqrt{2}-3)} = \frac{\sqrt{6}-3\sqrt{3}}{2-9} = \frac{\sqrt{6}-3\sqrt{3}}{-7} = \frac{3\sqrt{3}-\sqrt{6}}{7}.$$

And rationalizing the denominator of $\dfrac{a\sqrt{b}+c\sqrt{d}}{x\sqrt{y}+u\sqrt{v}}$, we can get:

$$\frac{a\sqrt{b}+c\sqrt{d}}{x\sqrt{y}+u\sqrt{v}} = \frac{(a\sqrt{b}+c\sqrt{d})(x\sqrt{y}-u\sqrt{v})}{(x\sqrt{y}+u\sqrt{v})(x\sqrt{y}-u\sqrt{v})} = \frac{(a\sqrt{b}+c\sqrt{d})(x\sqrt{y}-u\sqrt{v})}{x^2 y - u^2 v}.$$

What then, about this: $\dfrac{\sqrt{5}}{\sqrt[3]{3}+2}$?

Using the identities $a^3 + b^3 = (a+b)(a^2 - ab + b^2)$ and $a^3 - b^3 = (a-b)(a^2 + ab + b^2)$, we can rationalize some denominators the way below:

$$\frac{1}{\sqrt[3]{a}+\sqrt[3]{b}} = \frac{\{(\sqrt[3]{a})^2 - \sqrt[3]{a}\sqrt[3]{b} + (\sqrt[3]{b})^2\}}{(\sqrt[3]{a}+\sqrt[3]{b})\{(\sqrt[3]{a})^2 - \sqrt[3]{a}\sqrt[3]{b} + (\sqrt[3]{b})^2\}} = \frac{\sqrt[3]{a^2} - \sqrt[3]{ab} + \sqrt[3]{b^2}}{(\sqrt[3]{a}+\sqrt[3]{b})(\sqrt[3]{a^2} - \sqrt[3]{ab} + \sqrt[3]{b^2})}$$

$$= \frac{\sqrt[3]{a^2} - \sqrt[3]{ab} + \sqrt[3]{b^2}}{a+b}.$$

$$\frac{1}{\sqrt[3]{a}-\sqrt[3]{b}} = \frac{\{(\sqrt[3]{a})^2 + \sqrt[3]{a}\sqrt[3]{b} + (\sqrt[3]{b})^2\}}{(\sqrt[3]{a}+\sqrt[3]{b})\{(\sqrt[3]{a})^2 + \sqrt[3]{a}\sqrt[3]{b} + (\sqrt[3]{b})^2\}} = \frac{\sqrt[3]{a^2} + \sqrt[3]{ab} + \sqrt[3]{b^2}}{(\sqrt[3]{a}-\sqrt[3]{b})(\sqrt[3]{a^2} + \sqrt[3]{ab} + \sqrt[3]{b^2})}$$

$$= \frac{\sqrt[3]{a^2} + \sqrt[3]{ab} + \sqrt[3]{b^2}}{a-b}.$$

So in sum, we can get:

$$\frac{1}{\sqrt[3]{a}+\sqrt[3]{b}}=\frac{\sqrt[3]{a^2}-\sqrt[3]{ab}+\sqrt[3]{b^2}}{a+b}, \text{ and } \frac{1}{\sqrt[3]{a}-\sqrt[3]{b}}=\frac{\sqrt[3]{a^2}+\sqrt[3]{ab}+\sqrt[3]{b^2}}{a-b}.$$

And thus, getting back to $\dfrac{\sqrt{5}}{\sqrt[3]{3}+2}$, and rationalizing the denominator, we can get:

$$\frac{\sqrt{5}}{\sqrt[3]{3}+2}=\frac{\sqrt{5}\{(\sqrt[3]{3})^2-2\sqrt[3]{3}+2^2\}}{(\sqrt[3]{3}+2)\{(\sqrt[3]{3})^2-2\sqrt[3]{3}+2^2\}}=\frac{\sqrt{5}(\sqrt[3]{9}-2\sqrt[3]{3}+4)}{(\sqrt[3]{3}+2)(\sqrt[3]{9}-2\sqrt[3]{3}+4)}$$

$$=\frac{\sqrt{5}(\sqrt[3]{9}-2\sqrt[3]{3}+4)}{(\sqrt[3]{3})^3+2^3}=\frac{\sqrt{5}(\sqrt[3]{9}-2\sqrt[3]{3}+4)}{3+8}=\frac{\sqrt{5}(\sqrt[3]{9}-2\sqrt[3]{3}+4)}{11}.$$

What then, about this: $\dfrac{\sqrt{5}}{2\sqrt[3]{3}+2}$?

$$\frac{\sqrt{5}}{2\sqrt[3]{3}+2}=\frac{\sqrt{5}}{2(\sqrt[3]{3}+1)}=\frac{\sqrt{5}\{(\sqrt[3]{3})^2-1\sqrt[3]{3}+1^2\}}{2(\sqrt[3]{3}+1)\{(\sqrt[3]{3})^2-1\sqrt[3]{3}+1^2\}}=\frac{\sqrt{5}(\sqrt[3]{9}-\sqrt[3]{3}+1)}{2\{(\sqrt[3]{3})^3+1^3\}}$$

$$=\frac{\sqrt{5}(\sqrt[3]{9}-\sqrt[3]{3}+1)}{2(3+1)}=\frac{\sqrt{5}(\sqrt[3]{9}-\sqrt[3]{3}+1)}{8}.$$

And for another instance, rationalizing the denominator of $\dfrac{\sqrt{5}}{2\sqrt[3]{3}+\sqrt[3]{2}}$, we can get:

$$\frac{\sqrt{5}}{2\sqrt[3]{3}+\sqrt[3]{2}}=\frac{\sqrt{5}}{\sqrt[3]{8}\sqrt[3]{3}+\sqrt[3]{2}}=\frac{\sqrt{5}}{\sqrt[3]{8\cdot3}+\sqrt[3]{2}}=\frac{\sqrt{5}}{\sqrt[3]{24}+\sqrt[3]{2}}$$

$$=\frac{\sqrt{5}(\sqrt[3]{24^2}-\sqrt[3]{24\cdot2}+\sqrt[3]{2^2})}{(\sqrt[3]{24}+\sqrt[3]{2})(\sqrt[3]{24^2}-\sqrt[3]{24\cdot2}+\sqrt[3]{2^2})}=\frac{\sqrt{5}(\sqrt[3]{24^2}-\sqrt[3]{48}+\sqrt[3]{2^2})}{24+2}$$

Meanwhile, we can get:

$$\sqrt[3]{24^2} - \sqrt[3]{48} + \sqrt[3]{2^2} = \sqrt[3]{(8 \cdot 3)^2} - \sqrt[3]{6 \cdot 8} + \sqrt[3]{4} = \sqrt[3]{8^2}\sqrt[3]{3^2} - \sqrt[3]{8}\sqrt[3]{6} + \sqrt[3]{4}$$

$$= (\sqrt[3]{8})^2 \cdot \sqrt[3]{9} - 2\sqrt[3]{6} + \sqrt[3]{4} = 4\sqrt[3]{9} - 2\sqrt[3]{6} + \sqrt[3]{4}.$$

So we get: $\dfrac{\sqrt{5}}{2\sqrt[3]{3} + \sqrt[3]{2}} = \dfrac{\sqrt{5}(\sqrt[3]{24^2} - \sqrt[3]{48} + \sqrt[3]{2^2})}{24 + 2} = \dfrac{\sqrt{5}(4\sqrt[3]{9} - 2\sqrt[3]{6} + \sqrt[3]{4})}{26}$.

And also, using the identity $(x + y)^2 = x^2 + 2xy + y^2$ or $(x - y)^2 = x^2 - 2xy + y^2$, we can simplify some irrational numbers the way below:

$$\sqrt{a + b + 2\sqrt{ab}} = \sqrt{a} + \sqrt{b}, \text{ because we have: } (\sqrt{a} + \sqrt{b})^2 = a + 2\sqrt{ab} + b.$$

And by the same token, we can have: $\sqrt{a + b - 2\sqrt{ab}} = \sqrt{a} - \sqrt{b}$ if $a \geq b$.

If however, $a < b$, we get: $\sqrt{a + b - 2\sqrt{ab}} = \sqrt{b} - \sqrt{a}$, simply because $\sqrt{a + b - 2\sqrt{ab}} > 0$.

For instance, we get: $\sqrt{5 - 2\sqrt{6}} = \sqrt{2 + 3 - 2\sqrt{6}} = \sqrt{3} - \sqrt{2}$. That's because we can get:

$$(\sqrt{3} - \sqrt{2})^2 = 3 - 2\sqrt{3}\sqrt{2} + 2 = 3 + 2 - 2\sqrt{3 \cdot 2} = 5 - 2\sqrt{6} \Rightarrow (\sqrt{3} - \sqrt{2})^2 = 5 - 2\sqrt{6}$$

$$\Rightarrow \sqrt{5 - 2\sqrt{6}} = \sqrt{(\sqrt{3} - \sqrt{2})^2} = \sqrt{3} - \sqrt{2} \Rightarrow \sqrt{5 - 2\sqrt{6}} = \sqrt{3} - \sqrt{2}.$$

Examples 4 in Irrationals

Simplify each radical below or put each in a mixed radical if it can be done so. A mixed radical is a radical multiplied by a rational such as $2\sqrt{3}$ and $4\sqrt[3]{2}$.

0. $\sqrt{\frac{1}{4}}$ 1. $\sqrt{\frac{1}{8}}$ 2. $\sqrt{\frac{-3}{8}}$ 3. $\sqrt[3]{\frac{8}{-27}}$

4. $\sqrt[3]{-\frac{8}{49}}$ 5. $\sqrt{\frac{7}{8}}$ 6. $\sqrt{\frac{3}{-4}}$ 7. $\sqrt[3]{\frac{2}{27}}$

8. $\sqrt[3]{-\frac{16}{27}}$ 9. $\sqrt{\frac{4}{27}}$ A. $\sqrt[3]{\frac{27}{4}}$ B. $\sqrt[3]{\frac{108}{27}}$

C. $\sqrt{\frac{45}{98}}$ D. $\sqrt[3]{\frac{1}{4}}$ E. $\sqrt[3]{\frac{1}{24}}$ F. $\sqrt[3]{\frac{3}{8}}$

G. $\sqrt[3]{\frac{4}{9}}$ H. $\sqrt{\frac{4}{9}}$ I. $\sqrt{\frac{2}{9}}$ J. $\sqrt{-\frac{4}{49}}$

K. $\sqrt[3]{\frac{15}{18}}$ L. $\sqrt[3]{\frac{-1}{24}}$ M. $\sqrt{\frac{8}{3}}$ N. $\sqrt[3]{\frac{8}{81}}$

O. $\sqrt[3]{\frac{16}{9}}$ P. $\sqrt[3]{\frac{135}{98}}$

Suggestions or Solutions
To the Examples 4 in Irrationals

0. $\sqrt{\frac{1}{4}} = \sqrt{(\frac{1}{2})^2} = \frac{1}{2}$, and we can put it this way, too, of course: $\sqrt{\frac{1}{4}} = \sqrt{\frac{1}{2^2}} = \frac{\sqrt{1}}{\sqrt{2^2}} = \frac{1}{2}$.

1. $\sqrt{\frac{1}{8}} = \sqrt{\frac{1}{4} \cdot \frac{1}{2}} = \sqrt{(\frac{1}{2})^2 \frac{1}{2}} = \frac{1}{2}\sqrt{\frac{1}{2}}$, and this way, too: $\sqrt{\frac{1}{8}} = \sqrt{(\frac{1}{2})^3} = \sqrt{(\frac{1}{2})^2 \frac{1}{2}} = \frac{1}{2}\sqrt{\frac{1}{2}}$.

2. $\sqrt{\frac{-3}{8}}$ cannot be made within the set of all real numbers, and thus, is not allowed, because no real number can be squared to be a negative number.

3. $\sqrt[3]{\frac{8}{-27}} = \sqrt[3]{-\frac{8}{27}} = -\sqrt[3]{\frac{2^3}{3^3}} = -\sqrt[3]{(\frac{2}{3})^3} = -\frac{2}{3}$

4. $\sqrt[3]{-\frac{8}{49}} = -\sqrt[3]{\frac{8}{49}} = -\sqrt[3]{\frac{2^3}{49}} = -\frac{\sqrt[3]{2^3}}{\sqrt[3]{49}} = -\frac{2}{\sqrt[3]{49}}$, which can be put the way below, too:

$-\frac{2}{\sqrt[3]{49}} = -\frac{2\sqrt[3]{7}}{\sqrt[3]{49}\sqrt[3]{7}} = -\frac{2\sqrt[3]{7}}{\sqrt[3]{49 \cdot 7}} = -\frac{2\sqrt[3]{7}}{\sqrt[3]{7^3}} = -\frac{2\sqrt[3]{7}}{7}$

5. $\sqrt{\frac{7}{8}} = \sqrt{\frac{7}{4 \cdot 2}} = \sqrt{\frac{1}{4} \cdot \frac{7}{2}} = \sqrt{(\frac{1}{2})^2 \frac{7}{2}} = \frac{1}{2}\sqrt{\frac{7}{2}}$

6. $\sqrt{\frac{3}{-4}}$ cannot be made within the set of all real numbers, and thus, is not allowed, because no real number can be squared to be a negative number.

7. $\sqrt[3]{\frac{2}{27}} = \sqrt[3]{\frac{2}{3^3}} = \sqrt[3]{\frac{1}{3^3} \cdot 2} = \sqrt[3]{(\frac{1}{3})^3 2} = \sqrt[3]{(\frac{1}{3})^3} \cdot \sqrt[3]{2} = \frac{1}{3}\sqrt[3]{2}$, which equals: $\frac{\sqrt[3]{2}}{3}$.

8. $\sqrt[3]{-\frac{16}{27}} = -\sqrt[3]{\frac{8 \cdot 2}{3^3}} = -\frac{1}{3}\sqrt[3]{8 \cdot 2} = -\frac{1}{3}\sqrt[3]{2^3 \cdot 2} = -\frac{1}{3}\sqrt[3]{2^3} \cdot \sqrt[3]{2} = -\frac{2}{3}\sqrt[3]{2}$, which equals: $-\frac{2\sqrt[3]{2}}{3}$.

9. $\sqrt{\frac{4}{27}} = \sqrt{\frac{2^2}{3^2 \cdot 3}} = \sqrt{\frac{2^2}{3^2} \cdot \frac{1}{3}} = \sqrt{(\frac{2}{3})^2 \frac{1}{3}} = \frac{2}{3}\sqrt{\frac{1}{3}} = \frac{2}{3} \cdot \frac{1}{\sqrt{3}} = \frac{2}{3\sqrt{3}} = \frac{2\sqrt{3}}{3\sqrt{3}\sqrt{3}} = \frac{2\sqrt{3}}{3 \cdot 3} = \frac{2\sqrt{3}}{9}$

A. $\sqrt[3]{\frac{27}{4}} = \sqrt[3]{\frac{3^3}{4}} = \frac{\sqrt[3]{3^3}}{\sqrt[3]{4}} = \frac{3}{\sqrt[3]{4}} = \frac{3\sqrt[3]{2}}{\sqrt[3]{4} \cdot \sqrt[3]{2}} = \frac{3\sqrt[3]{2}}{\sqrt[3]{4 \cdot 2}} = \frac{3\sqrt[3]{2}}{\sqrt[3]{2^3}} = \frac{3\sqrt[3]{2}}{2} = \frac{3}{2}\sqrt[3]{2}$

B. $\sqrt[3]{\frac{108}{27}} = \sqrt[3]{\frac{27 \cdot 4}{3^3}} = \sqrt[3]{\frac{3^3 \cdot 4}{3^3}} = \sqrt[3]{4}$

C. $\sqrt{\frac{45}{98}} = \sqrt{\frac{9 \cdot 5}{49 \cdot 2}} = \sqrt{\frac{3^2 \cdot 5}{7^2 \cdot 2}} = \sqrt{\frac{3^2}{7^2}}\sqrt{\frac{5}{2}} = \frac{3}{7}\sqrt{\frac{5}{2}} = \frac{3}{7}\frac{\sqrt{5}}{\sqrt{2}} = \frac{3}{7}\frac{\sqrt{5}\sqrt{2}}{\sqrt{2}\sqrt{2}} = \frac{3}{7}\frac{\sqrt{5 \cdot 2}}{2} = \frac{3\sqrt{10}}{14}$

D. $\sqrt[3]{\frac{1}{4}}$ does not really have to be simplified, but can be changed to $\sqrt[3]{\frac{2}{8}} = \frac{\sqrt[3]{2}}{\sqrt[3]{8}} = \frac{\sqrt[3]{2}}{2}$.

E. $\sqrt[3]{\frac{1}{24}} = \sqrt[3]{\frac{1}{8 \cdot 3}} = \sqrt[3]{\frac{1}{8}} \cdot \sqrt[3]{\frac{1}{3}} = \frac{1}{2}\sqrt[3]{\frac{1}{3}} = \frac{1}{2}\frac{1}{\sqrt[3]{3}} = \frac{1}{2\sqrt[3]{3}} = \frac{\sqrt[3]{3^2}}{2\sqrt[3]{3} \cdot \sqrt[3]{3^2}} = \frac{\sqrt[3]{3^2}}{2 \cdot 3} = \frac{\sqrt[3]{9}}{6}$

F. $\sqrt[3]{\frac{3}{8}} = \sqrt[3]{\frac{3}{2^3}} = \frac{\sqrt[3]{3}}{\sqrt[3]{2^3}} = \frac{\sqrt[3]{3}}{2}$

G. $\sqrt[3]{\frac{4}{9}} = \sqrt[3]{\frac{12}{27}} = \frac{\sqrt[3]{12}}{\sqrt[3]{27}} = \frac{\sqrt[3]{12}}{3}$

H. $\sqrt{\frac{4}{9}} = \sqrt{\left(\frac{2}{3}\right)^2} = \frac{2}{3}$

I. $\sqrt{\frac{2}{9}} = \frac{\sqrt{2}}{\sqrt{9}} = \frac{\sqrt{2}}{3}$

J. $\sqrt{-\frac{4}{49}}$ cannot be made within the set of all real numbers, and thus, is not allowed, because no real number can be squared to be a negative number.

K. $\sqrt[3]{\frac{15}{18}} = \sqrt[3]{\frac{5}{6}}$, which does not really have to be simplified, but can be changed to

$\sqrt[3]{\frac{5 \cdot 36}{6 \cdot 36}} = \frac{\sqrt[3]{180}}{6}$.

L. $\sqrt[3]{\frac{-1}{24}} = -\sqrt[3]{\frac{1}{24}} = -\sqrt[3]{\frac{1}{8 \cdot 3}} = -\sqrt[3]{\frac{1}{8}} \cdot \sqrt[3]{\frac{1}{3}} = -\frac{1}{2}\sqrt[3]{\frac{9}{27}} = -\frac{1}{2}\frac{\sqrt[3]{9}}{\sqrt[3]{27}} = -\frac{1}{2}\frac{\sqrt[3]{9}}{3} = -\frac{\sqrt[3]{9}}{6}$

M. $\sqrt{\frac{8}{3}} = \sqrt{\frac{4 \cdot 2}{3}} = 2\sqrt{\frac{2}{3}} = 2\frac{\sqrt{2}}{\sqrt{3}} = \frac{2\sqrt{2}}{\sqrt{3}} = \frac{2\sqrt{2}\sqrt{3}}{\sqrt{3}\sqrt{3}} = \frac{2\sqrt{6}}{3}$

N. $\sqrt[3]{\frac{8}{81}} = \frac{\sqrt[3]{8}}{\sqrt[3]{81}} = \frac{2\sqrt[3]{9}}{\sqrt[3]{81 \cdot 9}} = \frac{2\sqrt[3]{9}}{9}$

O. $\sqrt[3]{\frac{16}{9}} = \sqrt[3]{\frac{16 \cdot 3}{27}} = \frac{\sqrt[3]{48}}{\sqrt[3]{27}} = \frac{\sqrt[3]{48}}{3}$

P. $\sqrt[3]{\frac{135}{98}} = \sqrt[3]{\frac{27 \cdot 5}{49 \cdot 2}} = \sqrt[3]{\frac{3^3 \cdot 5 \cdot 7}{7^3 \cdot 2}} = \frac{3}{7}\sqrt[3]{\frac{5 \cdot 7}{2}} = \frac{3}{7}\sqrt[3]{\frac{5 \cdot 7 \cdot 4}{2^3}} = \frac{3}{14}\sqrt[3]{140} = \frac{3\sqrt[3]{140}}{14}$

Examples 5 in Irrationals

Simplify each radical below or put each in a mixed radical if it can be done so. A mixed radical is a radical multiplied by a rational such as $2\sqrt{3}$ and $4\sqrt[3]{2}$.

0. $\sqrt{2\sqrt{4}}$ 1. $\sqrt{2\sqrt{16}}$ 2. $\sqrt{3\sqrt{36}}$ 3. $\sqrt{6\sqrt{225}}$

4. $\sqrt{2\sqrt[3]{125}}$ 5. $\sqrt[3]{4\sqrt{9}}$ 6. $\sqrt[3]{4\sqrt{4}}$ 7. $\sqrt[3]{32\sqrt{16}}$

8. $\sqrt[3]{10\sqrt[3]{64}}$ 9. $\sqrt[3]{0.8}$ A. $\sqrt[3]{0.008}$ B. $\sqrt[3]{0.016}$

C. $\sqrt[3]{\frac{0.001}{0.027}}$ D. $\sqrt[3]{\frac{0.01}{2.7}}$ E. $\sqrt{\frac{4.84}{9}}$ F. $\sqrt{\frac{0.0004}{0.81}}$

G. $\sqrt[3]{\frac{-0.001}{0.027}}$ H. $\sqrt{\frac{0.4}{0.009}}$ I. $\sqrt[3]{\frac{8}{0.081}}$ J. $\sqrt[3]{\frac{1.5}{9}}$

Suggestions or Solutions
To the Examples 5 in Irrationals

0. $\sqrt{2\sqrt{4}} = \sqrt{2\sqrt{2^2}} = \sqrt{2\cdot 2} = 2.$ So by the same token, we can have:

$$\sqrt{6\sqrt{3}\sqrt{4\sqrt{9}}} = \sqrt{6\sqrt{3}\sqrt{4\sqrt{3^2}}} = \sqrt{6\sqrt{3}\sqrt{4\cdot 3}} = \sqrt{6\sqrt{3}\sqrt{12}} = \sqrt{6\sqrt{36}} = \sqrt{6^2} = 6.$$

1. $\sqrt{2\sqrt{16}} = \sqrt{2\sqrt{4^2}} = \sqrt{2\cdot 4} = \sqrt{2}\sqrt{4} = 2\sqrt{2}$

2. $\sqrt{3\sqrt{36}} = \sqrt{3\cdot 6} = \sqrt{3\cdot 3\cdot 2} = 3\sqrt{2}$

3. $\sqrt{6\sqrt{225}} = \sqrt{6\sqrt{25\cdot 9}} = \sqrt{6\sqrt{15\cdot 15}} = \sqrt{6\cdot 15} = \sqrt{3\cdot 2\cdot 3\cdot 5} = 3\sqrt{10}$

4. $\sqrt{2\sqrt[3]{125}} = \sqrt{2\sqrt[3]{5^3}} = \sqrt{2\cdot 5} = \sqrt{10}$

5. $\sqrt[3]{4\sqrt{9}} = \sqrt[3]{4\cdot 3} = \sqrt[3]{12}$

6. $\sqrt[3]{4\sqrt{4}} = \sqrt[3]{4\cdot 2} = 2$

7. $\sqrt[3]{32\sqrt{16}} = \sqrt[3]{8\cdot 4\cdot 4} = \sqrt[3]{8\cdot 8\cdot 2} = \sqrt[3]{8}\cdot\sqrt[3]{8}\cdot\sqrt[3]{2} = 2\cdot 2\sqrt[3]{2} = 4\sqrt[3]{2}$

8. $\sqrt[3]{10\sqrt[3]{64}} = \sqrt[3]{10\sqrt[3]{4^3}} = \sqrt[3]{10\cdot 4} = \sqrt[3]{5\cdot 8} = 2\sqrt[3]{5}$

9. $\sqrt{0.8} = \sqrt{\frac{8}{10}} = \sqrt{\frac{4}{5}} = \frac{\sqrt{4}}{\sqrt{5}} = \frac{2}{\sqrt{5}} = \frac{2\sqrt{5}}{\sqrt{5}\sqrt{5}} = \frac{2\sqrt{5}}{5}$

A. $\sqrt[3]{0.008} = \sqrt[3]{0.001 \cdot 8} = \sqrt[3]{(0.1)^3 \cdot 8} = \sqrt[3]{(0.1)^3} \cdot \sqrt[3]{8} = 0.1 \cdot 2 = 0.2$.

And of course, we can put it this way, too: $\sqrt[3]{0.008} = \sqrt[3]{(0.2)^3} = 0.2$.

B. $\sqrt[3]{0.016} = \sqrt[3]{\frac{16}{1000}} = \sqrt[3]{\frac{8 \cdot 2}{1000}} = \sqrt[3]{\frac{8}{1000} \cdot 2} = \sqrt[3]{(\frac{2}{10})^3 \cdot 2} = \sqrt[3]{(\frac{2}{10})^3} \cdot \sqrt[3]{2} = \frac{2}{10} \cdot \sqrt[3]{2} = \frac{1}{5} \cdot \sqrt[3]{2} = \frac{\sqrt[3]{2}}{5}$

C. $\sqrt[3]{\frac{0.001}{0.027}} = \sqrt[3]{\frac{1}{27}} = \frac{1}{3}$. And we can put it this way, too: $\sqrt[3]{\frac{0.001}{0.027}} = \sqrt[3]{(\frac{0.1}{0.3})^3} = \sqrt[3]{(\frac{1}{3})^3} = \frac{1}{3}$.

D. $\sqrt[3]{\frac{0.01}{2.7}} = \sqrt[3]{\frac{0.1}{27}} = \sqrt[3]{\frac{1}{27} \cdot 0.1} = \sqrt[3]{\frac{1}{27}} \cdot \sqrt[3]{0.1} = \frac{1}{3}\sqrt[3]{\frac{1}{10}} = \frac{1}{3} \cdot \frac{1}{\sqrt[3]{10}} = \frac{1}{3} \cdot \frac{\sqrt[3]{100}}{\sqrt[3]{10} \cdot \sqrt[3]{100}} = \frac{1}{3} \cdot \frac{\sqrt[3]{100}}{10} = \frac{\sqrt[3]{100}}{30}$

E. $\sqrt{\frac{4.84}{9}} = \sqrt{\frac{484}{900}} = \sqrt{\frac{121 \cdot 4}{900}} = \sqrt{\frac{121 \cdot 4}{225 \cdot 4}} = \sqrt{\frac{121}{225}} = \sqrt{(\frac{11}{15})^2} = \frac{11}{15}$

F. $\sqrt{\frac{0.0004}{0.81}} = \sqrt{\frac{4}{8100}} = \sqrt{(\frac{2}{90})^2} = \frac{2}{90} = \frac{1}{45}$

G. $\sqrt[3]{\frac{-0.001}{0.027}} = -\sqrt[3]{\frac{1}{27}} = -\frac{1}{3}$

H. $\sqrt{\frac{0.4}{0.009}} = \sqrt{\frac{400}{9}} = \sqrt{(\frac{20}{3})^2} = \frac{20}{3}$

I. $\sqrt[3]{\frac{8}{0.081}} = \sqrt[3]{\frac{8000}{81}} = \sqrt[3]{\frac{20^3}{27 \cdot 3}} = \sqrt[3]{\frac{20^3}{3^3} \cdot \frac{1}{3}} = \frac{20}{3}\sqrt[3]{\frac{1}{3}} = \frac{20}{3} \cdot \frac{1}{\sqrt[3]{3}} = \frac{20}{3} \cdot \frac{\sqrt[3]{27}}{\sqrt[3]{3} \cdot \sqrt[3]{27}} = \frac{20}{3} \cdot \frac{\sqrt[3]{27}}{3} = \frac{20\sqrt[3]{27}}{9}$

J. $\sqrt[3]{\frac{1.5}{9}} = \sqrt[3]{\frac{15}{90}} = \sqrt[3]{\frac{3}{18}} = \sqrt[3]{\frac{1}{6}} = \frac{1}{\sqrt[3]{6}} = \frac{\sqrt[3]{36}}{\sqrt[3]{6} \cdot \sqrt[3]{36}} = \frac{\sqrt[3]{36}}{6}$

Examples 6 in Irrationals

Simplify each radical below or put each in a mixed radical if it can be done so. A mixed radical is a radical multiplied by a rational such as $2\sqrt{3}$ and $4\sqrt[3]{2}$.

0. $\sqrt{x^2}$

1. $\sqrt{x^3}$

2. $\sqrt{x^4}$

3. $\sqrt{x^5}$

4. $\sqrt{xy^2}$

5. $\sqrt{x^2 y^2}$

6. $\sqrt{x^5 y^2}$

7. $\sqrt{y^6}$

8. $\sqrt{x^5 y^7}$

9. $\sqrt{|x|^2}$

A. $\sqrt{|x|^2 y^2}$

B. $\sqrt[3]{x^3}$

C. $\sqrt[3]{x^4}$

D. $\sqrt[3]{x^5}$

E. $\sqrt[3]{x^7}$

F. $\sqrt[3]{-x^{23}}$

G. $\sqrt[3]{-x^3 y^3}$

H. $\sqrt[3]{x^4 y^4}$

I. $\sqrt[3]{-x^4 y^4}$

J. $\sqrt{\frac{1}{x^2}}$

K. $\sqrt{\frac{1}{x^3}}$

L. $\sqrt{\frac{-2}{x^3}}$

M. $\sqrt[3]{\frac{8}{-x^3}}$

N. $\sqrt[3]{-\frac{4y^3}{x^6}}$

O. $\sqrt{\frac{7y^3}{8x^2}}$

P. $\sqrt{\frac{3x^2}{-4y}}$

Q. $\sqrt[3]{\frac{64y^2}{9x^3}}$

R. $\sqrt[3]{-\frac{16y^3 z^4}{81x^4}}$

S. $\sqrt{\frac{4y^3 z^3}{27x^3}}$

T. $\sqrt[3]{\frac{27z^5}{4x^4 y^5}}$

Suggestions or Solutions
To the Examples 6 in Irrationals

0. $\sqrt{x^2} = x$ if $x \geq 0$. If however, $x < 0$, we get: $\sqrt{x^2} = -x$, because $\sqrt{x^2} > 0$.

So in sum, we get: $\sqrt{x^2} = |x|$. What do we mean by $|x|$ though?

$|x|$ is the absolute value of, that is, the magnitude of x.
So for instance, we have: $|-2| = 2$, $|0| = 0$, and $|3| = 3$.

And we want to keep in mind that not only what's inside a square root sign but the square root, too, is ≥ 0.

So if for instance, $r = \sqrt{t}$, we get: $r \geq 0$, that is, $\sqrt{t} \geq 0$ as well as $t \geq 0$.

1. $\sqrt{x^3} = \sqrt{x^2 x} = x\sqrt{x}$ for $x \geq 0$.   Why not $\sqrt{x^2 x} = |x|\sqrt{x}$, though?

It is OK, too, to set: $\sqrt{x^2 x} = |x|\sqrt{x}$ for $x \geq 0$, because if $x \geq 0$, $|x|\sqrt{x} = x\sqrt{x}$.

And in $\sqrt{x^3}$, x cannot be negative, because what's inside a square root sign cannot be negative, that is, it can only be positive or 0.

So it is necessary to specify $x \geq 0$ in this case.

Just setting: $\sqrt{x^3} = \sqrt{x^2 x} = x\sqrt{x}$, we mean x can be any of all real numbers including negative numbers, that is, it can be negative as well as positive or 0.

2. $\sqrt{x^4} = \sqrt{(x^2)^2} = x^2$. Why not $|x^2|$ though?

We have: $|x^2| = x^2$, since $|x^2| \geq 0$, and so is x^2. So the absolute value sign is unnecessary.

3. $\sqrt{x^5} = \sqrt{x^4 x} = x^2 \sqrt{x}$ for $x \geq 0$.

4. $\sqrt{xy^2} = \sqrt{y^2} \sqrt{x} = |y| \sqrt{x}$ for $x \geq 0$.

5. $\sqrt{x^2 y^2} = \sqrt{x^2} \sqrt{y^2} = |x| \cdot |y| = |xy|$.

Note however, it is ***not always*** the case where $|x - y| = |x| - |y|$. So for instance, we have:

$3 - 2| = |1| = 1$, and $|3| - |2| = 3 - 2 = 1$.

$2 - 3| = |-1| = 1$, but $|2| - |3| = 2 - 3 = -1$.

So we have: $|x - y| \neq |x| - |y|$.

6. $\sqrt{x^5 y^2} = \sqrt{y^2} \sqrt{x^5} = |y| \sqrt{x^5} = |y| x^2 \sqrt{x}$ for $x \geq 0$.

7. $\sqrt{y^6} = \sqrt{(y^3)^2} = |y^3|$. That's because:

If $y \geq 0$, we get $y^3 \geq 0$. So it is the case where $\sqrt{y^6} = y^3$, because $\sqrt{y^6} \geq 0$.

If however, $y < 0$, we get: $y^3 < 0$, so we cannot just set: $\sqrt{y^6} = y^3$, because $\sqrt{y^6} > 0$.

And thus, we get: $\sqrt{y^6} = |y^3|$.

• So we want to be careful with the sign of the number or the variable when we take it out of a square root sign.

8. $\sqrt{x^5 y^7} = \sqrt{x^5}\sqrt{y^7} = \sqrt{x^4 x}\sqrt{y^6 y} = \sqrt{x^4}\sqrt{x}\sqrt{y^6}\sqrt{y} = x^2\sqrt{x} \cdot y^3\sqrt{y}$

$= x^2 y^3 \sqrt{x}\sqrt{y} = x^2 y^3 \sqrt{xy}$ for x and y both ≥ 0.

Why not $\sqrt{y^6}\sqrt{y} = |y^3|\sqrt{y}$, though?

Just setting: $\sqrt{y^6}\sqrt{y} = |y^3|\sqrt{y}$, we mean y can be any of all real numbers including negative numbers, that is, it can be negative as well as positive or 0.

And we can get the same result the way below, too:

$\sqrt{x^5 y^7} = \sqrt{x^4 y^6 xy} = \sqrt{(x^2 y^3)^2 xy} = x^2 y^3 \sqrt{xy}$ for x and y both ≥ 0.

9. $\sqrt{|x|^2}$ is nothing but $\sqrt{x^2}$, so we get: $\sqrt{|x|^2} = |x|$.

A. $\sqrt{|x|^2 y^2}$ is nothing but $\sqrt{x^2 y^2}$, so we get: $\sqrt{|x|^2 y^2} = \sqrt{x^2 y^2} = |xy|$.

B. $\sqrt[3]{x^3} = x$

C. $\sqrt[3]{x^4} = \sqrt[3]{x^3 x} = \sqrt[3]{x^3} \cdot \sqrt[3]{x} = x\sqrt[3]{x}$

D. $\sqrt[3]{x^5} = \sqrt[3]{x^3} \cdot \sqrt[3]{x^2} = x\sqrt[3]{x^2}$

E. $\sqrt[3]{x^7} = \sqrt[3]{x^6 x} = \sqrt[3]{x^6} \cdot \sqrt[3]{x} = \sqrt[3]{(x^2)^3} \cdot \sqrt[3]{x} = x^2 \sqrt[3]{x}$

F. $\sqrt[3]{-x^{23}} = -\sqrt[3]{x^{21} x^2} = -\sqrt[3]{x^{21}} \cdot \sqrt[3]{x^2} = -\sqrt[3]{(x^7)^3} \cdot \sqrt[3]{x^2} = -x^7 \sqrt[3]{x^2}$

G. $\sqrt[3]{-x^3 y^3} = -\sqrt[3]{x^3 y^3} = -\sqrt[3]{(xy)^3} = -xy$

H. $\sqrt[3]{x^4 y^4} = \sqrt[3]{x^3 y^3 xy} = \sqrt[3]{x^3 y^3} \cdot \sqrt[3]{xy} = \sqrt[3]{(xy)^3} \cdot \sqrt[3]{xy} = xy \sqrt[3]{xy}$

I. $\sqrt[3]{-x^4 y^4} = -\sqrt[3]{x^4 y^4} = -xy \sqrt[3]{xy}$

J. $\sqrt{\frac{1}{x^2}} = \frac{1}{\sqrt{x^2}} = \frac{1}{|x|} = \left|\frac{1}{x}\right|$ for $x \neq 0$.

K. $\sqrt{\frac{1}{x^3}} = \frac{1}{\sqrt{x^3}} = \frac{1}{\sqrt{x^2 x}} = \frac{1}{\sqrt{x^2}\sqrt{x}} = \frac{1}{x\sqrt{x}} = \frac{\sqrt{x}}{x\sqrt{x}\sqrt{x}} = \frac{\sqrt{x}}{x^2}$ for $x > 0$.

And we can put it the way below, too: $\sqrt{\frac{1}{x^3}} = \sqrt{\frac{x}{x^3 x}} = \frac{\sqrt{x}}{\sqrt{x^4}} = \frac{\sqrt{x}}{x^2}$ for $x > 0$.

L. $\sqrt{\frac{-2}{x^3}} = \sqrt{\frac{-2}{x} \cdot \frac{1}{x^2}} = \sqrt{\frac{-2}{x}} \sqrt{\frac{1}{x^2}} = \sqrt{\frac{-2}{x}} \sqrt{\left(\frac{1}{x}\right)^2} = -\frac{1}{x}\sqrt{\frac{-2}{x}}$ for $x < 0$. How com?

What's inside a square root sign is ≥ 0, which means in this case, x has to be < 0, and it cannot be 0, because no denominator can be 0. So x can only be negative in this case.

M. $\sqrt[3]{\frac{8}{-x^3}} = -\sqrt[3]{\left(\frac{2}{x}\right)^3} = -\frac{2}{x}$ for $x \neq 0$.

N. $\sqrt[3]{-\dfrac{4y^3}{x^6}} = -\sqrt[3]{4\cdot\dfrac{y^3}{x^6}} = -\sqrt[3]{4}\cdot\sqrt[3]{\dfrac{y^3}{x^6}} = -\sqrt[3]{4}\cdot\sqrt[3]{(\dfrac{y}{x^2})^3} = -\sqrt[3]{4}\cdot\dfrac{y}{x^2} = -\dfrac{y\sqrt[3]{4}}{x^2}$ for $x \neq 0$.

O. $\sqrt{\dfrac{7y^3}{8x^2}} = \dfrac{\sqrt{7y^3}}{\sqrt{8x^2}} = \dfrac{\sqrt{7y^2 y}}{\sqrt{4\cdot 2x^2}} = \dfrac{y\sqrt{7y}}{2|x|\sqrt{2}} = \dfrac{y\sqrt{7y}\sqrt{2}}{2|x|\sqrt{2}\sqrt{2}} = \dfrac{y\sqrt{14y}}{4|x|}$ for $x \neq 0$, and $y \geq 0$.

Note that $4|x| = |4x|$.

P. $\sqrt{\dfrac{3x^2}{-4y}} = \sqrt{\dfrac{x^2}{4}}\sqrt{\dfrac{3}{-y}} = \dfrac{|x|}{2}\sqrt{\dfrac{3}{-y}} = \dfrac{|x|}{2}\sqrt{\dfrac{-3y}{yy}} = \dfrac{|x|}{2}\sqrt{\dfrac{1}{y^2}}\sqrt{-3y} = \dfrac{|x|}{-2y}\sqrt{-3y}$ for $y < 0$.

Note that in the case above, y can only be negative, because no denominator can be 0, and what's inside a square root sign is ≥ 0.

Q. $\sqrt[3]{\dfrac{64y^2}{9x^3}} = \sqrt[3]{\dfrac{4^3}{x^3}}\cdot\sqrt[3]{\dfrac{y^2}{3^2}} = \dfrac{4}{x}\sqrt[3]{\dfrac{3y^2}{3^3 3}} = \dfrac{4}{3x}\sqrt[3]{3y^2} = \dfrac{4\sqrt[3]{3y^2}}{3x}$ for $x \neq 0$.

R. $\sqrt[3]{-\dfrac{16y^3 z^4}{81x^4}} = -\sqrt[3]{\dfrac{8\cdot 2y^3 z^3 z}{27\cdot 3x^3 x}} = -\sqrt[3]{\dfrac{(2yz)^3 2z}{(3x)^3 3x}} = -\sqrt[3]{\dfrac{(2yz)^3}{(3x)^3}}\cdot\sqrt[3]{\dfrac{2z}{3x}} = -\dfrac{2yz}{3x}\sqrt[3]{\dfrac{2z}{3x}}$ for $x \neq 0$.

S. $\sqrt{\dfrac{4y^3 z^3}{27x^3}} = \sqrt{\dfrac{2^2 y^3 z^3}{3^3 x^3}} = \sqrt{\dfrac{(2yz)^2 yz}{(3x)^2 3x}} = \dfrac{2yz}{3x}\sqrt{\dfrac{yz}{3x}} = \dfrac{2yz}{3x}\sqrt{\dfrac{yz3x}{3^2 x^2}} = \dfrac{2yz}{3x}\dfrac{\sqrt{3xyz}}{3x} = \dfrac{2yz\sqrt{3xyz}}{9x^2}$ for $x > 0$, and

y and z both ≥ 0.

T. $\sqrt[3]{\dfrac{27z^5}{4x^4 y^5}} = \sqrt[3]{\dfrac{2\cdot 3^3 z^3 z^2}{8x^3 y^3 xy^2}} = \sqrt[3]{\dfrac{3^3 z^3}{2^3 x^3 y^3}}\cdot\sqrt[3]{\dfrac{2z^2}{xy^2}} = \dfrac{3z}{2xy}\sqrt[3]{\dfrac{2z^2}{xy^2}} = \dfrac{3z}{2xy}\sqrt[3]{\dfrac{2yz^2}{xy^3}} = \dfrac{3z}{2xy^2}\sqrt[3]{\dfrac{2yz^2}{x}}$

$= \dfrac{3z}{2xy^2}\sqrt[3]{\dfrac{2x^2 yz^2}{x^3}} = \dfrac{3z}{2x^2 y^2}\sqrt[3]{2x^2 yz^2} = \dfrac{3z\sqrt[3]{2x^2 yz^2}}{2x^2 y^2}$ for x and y both $\neq 0$.

Examples 7 in Irrationals

Simplify each expression below.

0. $\sqrt{(2-\sqrt{5})^2}$ 1. $\sqrt{(\sqrt{3}-1)^2}$ 2. $\sqrt{(x^2-2x+4)^2}$

3. $\sqrt{(x^2-3x+2)^2}$ 4. $\sqrt{(3x-2-x^2)^2}$ 5. $\sqrt{(2x-4-x^2)^2}$

6. $\sqrt{x^2-2xy+y^2}$ 7. $\sqrt{(x-2)^2}+|2-x|$

Suggestions or Solutions
To the Examples 7 in Irrationals

0.   Simplify $\sqrt{(2-\sqrt{5})^2}$

We know $\sqrt{2^2} = 2$, so assumig: $N = \sqrt{(2-\sqrt{5})^2}$, do we just get: $N = 2 - \sqrt{5}$?

First, we have: $5 > 4 \Rightarrow \sqrt{5} > \sqrt{4} = 2 \Rightarrow \sqrt{5} > 2 \Rightarrow 2 - \sqrt{5} < 0$.

Next, taking a square root of a number, we can only get a number positive or 0. And the number is the square root. That is to say that the square root is positive or 0.

So we cannot have: $N < 0$. That is, $N \neq 2 - \sqrt{5}$. What then, is N?

We have: $N = \sqrt{(2-\sqrt{5})^2} = \sqrt{(\sqrt{5}-2)^2}$.

And we know: $\sqrt{5} - 2 > 0$. So we get: $N = \sqrt{5} - 2$.

Note that we have in fact, a fact that: $\sqrt[n]{k} \geq 0$ for any $k \geq 0$ if n is even.

If however, n is odd, we get: $\sqrt[n]{k} < 0$ if $k < 0$.

For instance, we have: $\sqrt[3]{-3} < 0$, and $\sqrt[8]{2} > 0$.

1. **Simplify** $\sqrt{(\sqrt{3}-1)^2}$

We have: $3 > 1 \Rightarrow \sqrt{3} > \sqrt{1} = 1 \Rightarrow \sqrt{3} > 1 \Rightarrow \sqrt{3} - 1 > 0$. So we get: $\sqrt{(\sqrt{3}-1)^2} = \sqrt{3} - 1$.

0.2. **Simplify** $\sqrt{(x^2 - 2x + 4)^2}$

Assuming first, $N = \sqrt{(x^2 - 2x + 4)^2}$, we need to have this first: $N \geq 0$.

That is, it is not always the case where we just get: $N = x^2 - 2x + 4$.

Thus next, we want check to see if $(x^2 - 2x + 4)$ can be negative.

So dong the check, we get: $x^2 - 2x + 4 = x^2 - 2x + 1 + 3 = (x-1)^2 + 3 > 0$ for all x.

And thus, we can now say that: $N = x^2 - 2x + 4$.

0.3. $\sqrt{(x^2 - 3x + 2)^2}$

Assuming first, $N = \sqrt{(x^2 - 3x + 2)^2}$, we can say for now, that $N \geq 0$.

Thus next, we want to check to see if $(x^2 - 3x + 2)$ can be negative.

We can get: $x^2 - 3x + 2 = (x-1)(x-2) \geq 0 \Rightarrow x \leq 1$ or $x \geq 2$.
That is to say that we get: $x^2 - 3x + 2 < 0$ if $1 < x < 2$.

So we get: $N = x^2 - 3x + 2$ for $x \leq 1$ or $x \geq 2$, or $N = -(x^2 - 3x + 2)$ for $1 < x < 2$.

And we can put it this way, too: $N = |x^2 - 3x + 2|$.

That is to say that we can set: $\sqrt{(x^2 - 3x + 2)^2} = |x^2 - 3x + 2|$.

0.4. $\sqrt{(3x-2-x^2)^2}$

Assuming first, $N = \sqrt{(3x-2-x^2)^2}$, we can say for now, that $N \geq 0$.

Thus next, we want to check to see if $(3x - 2 - x^2)$ can be negative.

We can get: $3x - 2 - x^2 = (1-x)(x-2) \geq 0 \Rightarrow (x-1)(x-2) \leq 0 \Rightarrow 1 \leq x \leq 2$.
That is, we get: $3x - 2 - x^2 < 0$ if $x < 1$ or $x > 2$.

So we get: $N = 3x - 2 - x^2$ for $1 \leq x \leq 2$, or $N = x^2 - 3x + 2$ for $x < 1$ or $x > 2$.

And we can put it this way, too: $N = |3x - 2 - x^2|$.

That is, we can set: $\sqrt{(3x-2-x^2)^2} = |3x - 2 - x^2| = |x^2 - 3x + 2|$.

0.5. $\sqrt{(2x-4-x^2)^2}$

Assuming first, $N = \sqrt{(2x-4-x^2)^2}$, we can say for now, that $N \geq 0$.

So next, we want to check to see if $(2x - 4 - x^2)$ can be negative.

We can get: $2x - 4 - x^2 = -x^2 + 2x - 4 = -(x^2 - 2x + 4) = -(x^2 - 2x + 1 + 3)$
$= -\{(x-1)^2 + 3\} = -(x-1)^2 - 3 \leq -3$, which is negative for all x.

So we get: $N = -(2x - 4 - x^2) = x^2 - 2x + 4$.

And we can put it this way, too: $N = |2x - 4 - x^2|$.

So we can see that: $\sqrt{(2x-4-x^2)^2} = |2x - 4 - x^2| = x^2 + 2x - 4$.

0.6. $\sqrt{x^2 - 2xy + y^2}$

Assuming first, $N = \sqrt{x^2 - 2xy + y^2}$, we can say for now, that $N \geq 0$.

So next, we want to check to see if what's inside the square root can be negative.

We can get: $x^2 - 2xy + y^2 = (x - y)^2$. So we get:

$N = (x^2 - 2xy + y^2)^{1/2} = \{(x - y)^2\}^{1/2} = x - y$ for $x \geq y$, or $N = -(x - y) = y - x$ for $x < y$,

And we can put it this way, too: $N = |x - y| = |y - x|$.

That is, we can set: $\sqrt{x^2 - 2xy + y^2} = |x - y| = |y - x|$.

0.7. $\sqrt{(x - 2)^2} + |2 - x|$

Assuming first, $N = \sqrt{(x - 2)^2} + |2 - x|$, we can say for now, that $N \geq 0$.

And next, we get: $x - 2 \geq 0 \Rightarrow x \geq 2$. And we get: $x - 2 < 0 \Rightarrow x < 2$.

So assuming $M = \sqrt{(x - 2)^2}$, we get: $M = x - 2$ for $x \geq 2$, or $M = 2 - x$ for $x < 2$.

What then, about $|2 - x|$?

The two vertical bars are called absolute value sign.

And we get: $|y| \geq 0$ for all y. So for instance, $|-3| = 3$.

Thus first, we get: $2 - x \geq 0 \Rightarrow x \leq 2$. So we get: $|2 - x| = 2 - x$ if $x \leq 2$.

So next, if $x > 2$, we get: $|2 - x| = -(2 - x) = x - 2$.

Assuming thus, $K = |2 - x|$, we get: $K = x - 2$ for $x > 2$, or $K = 2 - x$ for $x \leq 2$.

And thus, we get:

$N = M + K = x - 2 - (2 - x) = 2x - 4$ if $x \geq 2$.

$N = M + K = 2 - x + (2 - x) = 4 - 2x$ if $x < 2$.

Examples 8 in Irrationals

0. Show that $\sqrt{2}$ is an irrational number.

1. Show that $\sqrt{3}$ is an irrational number.

2. Show that $\sqrt{6}$ is an irrational number.

3. Show that $\sqrt{2} + \sqrt{3}$ is an irrational number.

Suggestions or Solutions
To the Examples 8 in Irrationals

0.   Show that $\sqrt{2}$ is an irrational number.

A rational number can be put in a ratio between two integers prime to each other.

So assuming $\sqrt{2}$ is a rational number, we can put it this way: $\sqrt{2} = \dfrac{a}{b}$ where a and b are nonzero integers, and more importantly, are prime to each other.

Then, we can get: $2 = \dfrac{a^2}{b^2} \Rightarrow 2b^2 = a^2$.

In the problem 3 in the **Examples Y in Arithmetic 2** however, it was explained that if $a^2 = 2b^2$, a cannot be an integer if b is a nonzero integer. And vice versa.

So if $a^2 = 2b^2$, b cannot be an integer if a is a nonzero integer. And thus, we cannot get: $a^2 = 2b^2$ if a and b are nonzero integers, which contradicts the assumption above. So it is not the case where $\sqrt{2}$ is a rational number. That is, it's an irrational number.

Anyway, if $2b^2 = a^2$, we can say that a^2 is an integer multiple of 2, because b^2 is an integer, since b is an integer. So in short, a^2 is an even integer.

And thus, assuming k is an integer, we can set: $a^2 = 2k$, where a is an integer, of course. What integer then, is a?

It is an even integer, too.
We cannot get an even integer squaring an odd integer.
And we get an even integer squaring an even integer.
So assuming n is an integer, we can set: $a = 2n$.

Now, we have: $2b^2 = a^2$, and $a = 2n$. So we get: $2b^2 = 4n^2$, and thus, we get: $b^2 = 2n^2$.
So b^2 is an even integer. And thus, b is an even integer, too.
Assuming thus, m is an integer, we can set: $b = 2m$.   What then, do we get?

We get a contradiction.
It was mentioned that a and b are prime to each other.
However, we now have: $a = 2n$, and $b = 2m$.
So it is not the case where a and b are prime to each other.
And thus, the assumption that $\sqrt{2}$ is a rational number is not true.
So $\sqrt{2}$ is an irrational number.

1.   **Show that $\sqrt{3}$ is an irrational number.**

Suppose $\sqrt{3}$ is a rational number.

Then, we can put it this way: $\sqrt{3} = a/b$ where a and b are prime to each other.
And we can get: $3 = a^2/b^2 \Rightarrow 3b^2 = a^2$.

So we can say that a^2 is a multiple of 3.
And thus, assuming k is an integer, we can set: $a^2 = 3k$, where a is an integer, of course.
What integer then, is a?

It is a multiple of 3, too.
We cannot get a multiple of 3 squaring an integer that is not a multiple of 3.
Note that a multiple of 6, 9, or such is a multiple of 3.
And we get a multiple of 3 squaring an integer that is a multiple of 3.
So assuming n is an integer, we can set: $a = 3n$.

Now, we have: $3b^2 = a^2$, and $a = 3n$.

So we get: $3b^2 = 9n^2$, and thus, we get: $b^2 = 3n^2$.

So b^2 is a multiple of 3. And thus, b is a multiple of 3, too.

Assuming therefore, m is an integer, we can set: $b = 3m$.   What then, do we get?

We get a contradiction.

It was mentioned that a and b are prime to each other.

However, we now have: $a = 3n$, and $b = 3m$.

So it is not the case where a and b are prime to each other.

And thus, the assumption that $\sqrt{3}$ is a rational number is not true.

So it is not the case where $\sqrt{3}$ is a rational number.

And thus, $\sqrt{3}$ is an irrational number.

2.   Show that $\sqrt{6}$ is an irrational number.

Suppose $\sqrt{6}$ is a rational number.

Then, we can put it this way: $\sqrt{6} = a/b$ where a and b are prime to each other.

And we can get: $6 = a^2/b^2 \Rightarrow 6b^2 = a^2$.

So we can say that a^2 is a common multiple of 3 and 2, because it is a multiple of 6.

And thus, assuming k is an integer, we can set: $a^2 = 6k$, where a is an integer, of course.

What integer then, is a?

It is a multiple of 6.

We cannot get a multiple of 6 squaring an integer that is not a multiple of 6.

Note that a multiple of 12, 18, or such is a multiple of 6.

And we get a multiple of 6 squaring an integer that is a multiple of 6.

So assuming n is an integer, we can set: $a = 6n$.

Now, we have: $6b^2 = a^2$, and $a = 6n$.

So we get: $6b^2 = 36n^2$, and thus, we get: $b^2 = 6n^2$.

So b^2 is a multiple of 6. And thus, b is a multiple of 6, too.

Thus, assuming m is an integer, we can set: $b = 6m$.   What then, do we get?

We get a contradiction.

It was mentioned that a and b are prime to each other.

However, we now have: $a = 6n$, and $b = 6m$.

So it is not the case where a and b are prime to each other.

And thus, the assumption that $\sqrt{6}$ is a rational number is not true.

So $\sqrt{6}$ is an irrational number.

3.   **Show that $\sqrt{2} + \sqrt{3}$ is an irrational number.**

Suppose $\sqrt{2} + \sqrt{3} = r$, where r is a rational number.

Then first, we can put it this way: $\sqrt{2} = r - \sqrt{3}$.

So next, squaring both sides, we get:

$$2 = (r - \sqrt{3})^2 = r^2 - 2r\sqrt{3} + 3 \Rightarrow 2r\sqrt{3} = r^2 - 1 \Rightarrow \sqrt{3} = \frac{r^2-1}{2r}.$$

What then, do we get?

We get a contradiction.

We know $\sqrt{3}$ is proven irrational in the problem 1.2 above.

However, $\frac{r^2-1}{2r}$ is a rational number, because r is a rational number.

Doing arithmetic with rational numbers, we get a rational number.

So the assumption that $\sqrt{2} + \sqrt{3}$ is a rational number is not true.

And thus, $\sqrt{2} + \sqrt{3}$ is an irrational number.

Examples 9 in Irrationals

0. Simplify: $\sqrt{(x+\sqrt{x^2})^2} - \sqrt{(x-\sqrt{x^2})^2}$.

1. Assuming $x = 3a + b^3$, $y = 3b + a^3$, and $ab = 1$, find the value of the expression as follows: $\sqrt[3]{(x+y)^2} - \sqrt[3]{(x-y)^2}$.

2. Assuming $u \geq 0$, and $\sqrt{u} = 2a + 3$, simplify: $\sqrt{u + 4a + 7} - \sqrt{u - 36a + 27}$.

Suggestions or Solutions
To the Examples 9 in Irrationals

0. Simplify $\sqrt{(x+\sqrt{x^2})^2} - \sqrt{(x-\sqrt{x^2})^2}$

Assuming first, $x \geq 0$, we get: $\sqrt{x^2} = x$.

So we get: $\sqrt{(x+\sqrt{x^2})^2} - \sqrt{(x-\sqrt{x^2})^2} = \sqrt{(x+x)^2} - 0 = \sqrt{(2x)^2} = 2x$.

Assuming next, $x < 0$, we get: $\sqrt{x^2} = -x$.

So we get: $\sqrt{(x+\sqrt{x^2})^2} - \sqrt{(x-\sqrt{x^2})^2} = 0 - \sqrt{(x+x)^2} = -\sqrt{(2x)^2} = -(-2x) = 2x$.

Thus, we get: $\sqrt{(x+\sqrt{x^2})^2} - \sqrt{(x-\sqrt{x^2})^2} = 2x$.

So putting threads together, we get: $\sqrt{(x+\sqrt{x^2})^2} - \sqrt{(x-\sqrt{x^2})^2} = 2x$.

1. Assuming $x = 3a + b^3$, $y = 3b + a^3$, and $ab = 1$, find the value of the expression as follows: $\sqrt[3]{(x+y)^2} - \sqrt[3]{(x-y)^2}$. This is just for a algebra practice.

We are going to use a factorization identity $(a \pm b)^3 = a^3 \pm b^3 \pm 3ab(a + b)$.

$x + y = a^3 + b^3 + 3(a + b) = (a + b)^3 - 3ab(a + b) + 3(a + b) = (a + b)^3$, since $ab = 1$.

$x - y = b^3 - a^3 + 3(a - b) = (b - a)^3 + 3ab(b - a) - 3(b - a) = (b - a)^3$, since $ab = 1$.

So we get:

$$\sqrt[3]{(x+y)^2} - \sqrt[3]{(x-y)^2} = \sqrt[3]{(a+b)^6} - \sqrt[3]{(b-a)^6} = (a + b)^2 - (b - a)^2 = 4ab = 4.$$

2.   Assuming $u \geq 0$, and $\sqrt{u} = 2a + 3$, simplify: $\sqrt{u + 4a + 7} - \sqrt{u - 36a + 27}$.

This is just another practice for algebra.

We are given the expression $\sqrt{u} = 2a + 3$, so we want to use it, of course.

Looking at the expression to be simplified, u is a part of what's inside the radical sign. So it's a good idea to get u out of $\sqrt{u} = 2a + 3$. How?

Squaring both sides, we can get it.

So we get: $\sqrt{u} = 2a + 3 \Rightarrow u = (2a + 3)^2 = 4a^2 + 12a + 9$.

And next, how are we going to put it into the expression to be simplified?

The expression is made of two radicals, so we may want to take care of each at a time. And taking care of each, we want to take care of what's inside first.

So to begin with, we get:

$u + 4a + 7 = 4a^2 + 12a + 9 + 4a + 7 = 4a^2 + 16a + 16 = 4(a + 2)^2$.

And next, we get:

$u - 36a + 27 = 4a^2 + 12a + 9 - 36a + 27 = 4a^2 - 24a + 36 = 4(a - 3)^2$.

And thus, we now have: $\sqrt{u + 4a + 7} - \sqrt{u - 36a + 27} = \sqrt{4(a + 2)^2} - \sqrt{4(a - 3)^2}$.

So it seems we can rationalize the expression.

That is to say that we can now remove the radical signs.

We want to check the base though. What base?

What's inside the root sign is a power, which is made of a base and an exponent.

So in this case, $(a + 2)$ is the base in the power $(a + 2)^2$.

About what then, do we check the base?

We want to see if it can be negative or not.

That's because a square root of a real number is ≥ 0.

And in fact, both bases in both powers can be negative.

And we want to begin with the condition that $u \geq 0$.

And we have this, too: $\sqrt{u} = 2a + 3$.

Note that conditions are not given for nothing.

So to begin with, we can get: $u \geq 0 \Rightarrow 2a + 3 \geq 0 \Rightarrow a \geq -3/2$.

And checking with $(a + 2)$, we get: $a + 2 > 0$, since we have: $a \geq -3/2$.

What then, about $(a - 3)$?

If $a - 3 \geq 0$, we need to have: $a \geq 3$. However, we are given: $a \geq -3/2$.

That is to say that if $-3/2 \leq a < 3$, we get: $a - 3 < 0$.

So putting threads together, we have:

$-3/2 \leq a < 3 \Rightarrow a - 3 < 0$, but $a + 2 > 0$.

$a \geq 3 \Rightarrow a - 3 \geq 0$, and $a + 2 > 0$.

Thus first, we get: $-3/2 \leq a < 3 \Rightarrow \sqrt{4(a+2)^2} - \sqrt{4(a-3)^2}$

$= 2(a + 2) - \{-2(a - 3)\} = 2a + 4 + 2a - 6 = 4a - 2$.

And next, we get: $a \geq 3 \Rightarrow \sqrt{4(a+2)^2} - \sqrt{4(a-3)^2} = 2(a + 2) - 2(a - 3) = 10$.

Examples A in Irrationals

0.   Assuming a, b, x, and y are rationals, and C is an irrational, show that:

$a + bC = x + yC \Rightarrow a = x$, and $b = y$.

1.   Assuming a and b are rationals, and C is an irrational, show that

$a + bC = 0 \Rightarrow a = b = 0$.

2.   Assuming a and c are rationals, but B and D are irrationals, show that:

$a + B = c + D \Rightarrow a = c$, and $B = D$.

3.   Assuming a, b, and c are rationals, A, B, and C are irrationals, and $A \neq B \neq C$, show that: $aA + bB + cC = 0 \Rightarrow a = b = c = 0$.

Suggestions or Solutions
To the Examples A in Irrationals

0.   Assuming $a, b, x,$ and y are rationals, and C is an irrational, show that:
$a + bC = x + yC \Rightarrow a = x$, and $b = y$.

To begin with, we can put it this way: $a + bC = x + yC \Rightarrow a - x + (b - y)C = 0$.

Next, assuming $b - y \neq 0$, we can get: $C = \frac{a-x}{-(b-y)} = \frac{a-x}{y-b}$. What then, do we get?

We get a contradiction. How come?

We know $a, b, x,$ and y are rationals. So $a - x$ and $y - b$ are rationals, too.

So $\frac{a-x}{y-b}$ is a rational, too, because dividing a rational by another, we get a rational, too.

However, C is an irrational, which contradicts the fact that C is an irrational.

So it is not the case where $b - y \neq 0$, that is, we get: $b - y = 0 \Rightarrow b = y$.
What then, do we get?

We can get: $a - x + (b - y)C = 0 \Rightarrow a - x + 0 = 0 \Rightarrow a - x = 0 \Rightarrow a = x$.

1. Assuming a and b are rationals, and C is an irrational, show that
$a + bC = 0 \Rightarrow a = b = 0.$

Assuming first, $b \neq 0$, we can get: $C = -\dfrac{b}{a}$, which is a rational, which however, contradicts the fact that C is an irrational. So we get: $b = 0$.

So next, we get: $a + bC = 0 \Rightarrow a + 0 = 0 \Rightarrow a = 0$.

2. Assuming a and c are rationals, but B and D are irrationals, show that:
$a + B = c + D \Rightarrow a = c$, and $B = D$.

To begin with, $a - c$ is a rational.
Next, we can get: $a + B = c + D \Rightarrow a - c = D - B$, which is an irrational unless $D - B = 0$, since 0 is a rational. So what?

We know that $a - c$ has to be a rational, and cannot be irrational.
However, $D - B$ can be a rational if $D - B = 0$.

So if it is the case where $a - c = D - B$, we have to get: $D - B = 0$.
Thus, we get: $D - B = 0 \Rightarrow D = B$.

Then next, we get: $a - c = D - B = 0 \Rightarrow a - c = 0 \Rightarrow a = c$.

3. Assuming a, b, and c are rationals, A, B, and C are irrationals, and $A \neq B \neq C$, show that: $aA + bB + cC = 0 \Rightarrow a = b = c = 0$.

First, we can get: $aA + bB + cC = 0 \Rightarrow aA + bB = -cC \Rightarrow (aA + bB)^2 = (-cC)^2$
$\Rightarrow a^2A^2 + 2abAB + b^2B^2 = c^2C^2 \Rightarrow a^2A^2 + 2abAB + b^2B^2 - c^2C^2 = 0.$

Then, next, we can say that $a^2A^2 + b^2B^2 - c^2C^2$ can be a rational, but $2abAB$ cannot be a rational, and has to be an irrational, because $A \neq B$.

So from the example 1 above, we have to get: $a^2A^2 + b^2B^2 - c^2C^2 = 0$, and $ab = 0$.

And we get: $ab = 0 \Rightarrow a = 0$ or $b = 0$.

Then, first, we get:
$a = 0 \Rightarrow a^2A^2 + b^2B^2 - c^2C^2 = 0 \Rightarrow b^2B^2 - c^2C^2 = 0 \Rightarrow (bB - cC)(bB + cC)$
$\Rightarrow bB = cC$ or $bB = -cC$.   So we get:

$bB = cC \Rightarrow b = c\frac{C}{B} \Rightarrow b - c\frac{C}{B} = 0 \Rightarrow b = c = 0$ from the example 1, because $\frac{C}{B}$ is an irrational and b and c are rationals.

$bB = -cC \Rightarrow b = -c\frac{C}{B} \Rightarrow b + c\frac{C}{B} = 0 \Rightarrow b = c = 0$ from the example 1, because $\frac{C}{B}$ is an irrational and b and c are rationals.

So we get: $a = 0 \Rightarrow b = c = 0 \Rightarrow a = b = c = 0.$

Then, next, we get:
$b = 0 \Rightarrow a^2A^2 - c^2C^2 = 0 \Rightarrow a^2A^2 - c^2C^2 = 0 \Rightarrow (aA - cC)(aA + cC)$
$\Rightarrow aA = cC$ or $aA = -cC$.   So we get:

$aA = cC \Rightarrow a = c\frac{C}{A} \Rightarrow a - c\frac{C}{A} = 0 \Rightarrow a = c = 0$ from the example 1, because $\frac{C}{A}$ is an irrational and a and c are rationals.

$aA = -cC \Rightarrow a = -c\frac{C}{A} \Rightarrow a + c\frac{C}{A} = 0 \Rightarrow a = c = 0$ from the example 1, because $\frac{C}{A}$ is an irrational and a and c are rationals.

So we get: $b = 0 \Rightarrow a = c = 0 \Rightarrow a = b = c = 0.$

And thus, we get: $a = b = c = 0.$

Examples B in Irrationals

0. Assuming $a = \dfrac{\sqrt{2}-1}{\sqrt{2}+1}$, and $F(x) = x^3 - 6x^2$, find $F(a)$.

1. Assuming $p = 1 + \sqrt[3]{2} + \sqrt[3]{4}$, and $F(x) = x^3 - 3x^2 - 3x - 1$, find $F(p)$.

2. Assuming $q = 2 + \sqrt{3}$, and $F(p) = \dfrac{p^4 - p^3 - 9p^2 - 5p + 2}{p^2 - 4p + 3}$, find $F(q)$.

Suggestions or Solutions
To the Examples B in Irrationals

0. Assuming $a = \dfrac{\sqrt{2}-1}{\sqrt{2}+1}$, **and** $F(x) = x^3 - 6x^2$, **find** $F(a)$.

Simplifying a first, we can get: $a = \dfrac{\sqrt{2}-1}{\sqrt{2}+1} = \dfrac{(\sqrt{2}-1)^2}{(\sqrt{2}+1)(\sqrt{2}-1)} = \dfrac{2-2\sqrt{2}+1}{2-1} = 3 - 2\sqrt{2}.$

So we get: $2\sqrt{2} = 3 - a \Rightarrow 8 = 9 - 6a + a^2 \Rightarrow a^2 - 6a + 1 = 0.$

Next, we have: $F(x) = x^3 - 6x^2$, so we get: $F(a) = a^3 - 6a^2.$

We have: $a^2 - 6a + 1 = 0.$ So we get: $a^2 - 6a = -1 \Rightarrow a^3 - 6a^2 = -a = 2\sqrt{2} - 3.$

Thus, we get: $F(a) = 2\sqrt{2} - 3.$

And of course, since $f(a) = a^3 - 6a^2 = -a$, and $a = \dfrac{\sqrt{2}-1}{\sqrt{2}+1}$, we can put it this way, too:

$F(a) = \dfrac{1-\sqrt{2}}{\sqrt{2}+1}.$

1.   Assuming $p = 1 + \sqrt[3]{2} + \sqrt[3]{4}$, **and** $F(x) = x^3 - 3x^2 - 3x - 1$, **find** $F(p)$.

To begin with, we can get: $p - 1 = \sqrt[3]{2} + \sqrt[3]{4}$, so we get: $(p-1)^3 = (\sqrt[3]{2} + \sqrt[3]{4})^3.$

Meanwhile: $(p-1)^3 = p^3 - 3p^2 + 3p - 1$, and

$(\sqrt[3]{2} + \sqrt[3]{4})^3 = 2 + 3\sqrt[3]{2}\sqrt[3]{4}(\sqrt[3]{2} + \sqrt[3]{4}) + 4 = 6 + 3\sqrt[3]{8}(\sqrt[3]{2} + \sqrt[3]{4}) = 6 + 6(\sqrt[3]{2} + \sqrt[3]{4})$

$\Rightarrow (\sqrt[3]{2} + \sqrt[3]{4})^3 = 6(1 + \sqrt[3]{2} + \sqrt[3]{4}) = 6p$, since $p = 1 + \sqrt[3]{2} + \sqrt[3]{4}.$

So we get: $p^3 - 3p^2 + 3p - 1 = 6p \Rightarrow p^3 - 3p^2 - 3p - 1 = 0 \Rightarrow F(p) = 0.$

2. Assuming $q = 2 + \sqrt{3}$, and $F(p) = \dfrac{p^4 - p^3 - 9p^2 - 5p + 2}{p^2 - 4p + 3}$, find $F(q)$.

To begin with, we have: $q = 2 + \sqrt{3}$.

So we can get: $(q - 2)^2 = 3 \Rightarrow q^2 - 4q + 4 = 3 \Rightarrow q^2 - 4q + 1 = 0$.

Next, we have: $F(p) = \dfrac{p^4 - p^3 - 9p^2 - 5p + 2}{p^2 - 4p + 3}$.

So we can get: $F(q) = \dfrac{q^4 - q^3 - 9q^2 - 5q + 2}{q^2 - 4q + 3}$.

And doing a sythetic division of the numerator by $q^2 - 4q + 1$, we get :

$$
\begin{array}{r}
q^2 + 3q + 2 \\
\hline
q^2 - 4q + 1 \,\big|\, q^4 - q^3 - 9q^2 - 5q + 2 \\
q^4 - 4q^3 + q^2 \\
\hline
3q^3 - 10q^2 - 5q + 2 \\
3q^3 - 12q^2 + 3q \\
\hline
2q^2 - 8q + 2 \\
2q^2 - 8q + 2 \\
\hline
0
\end{array}
$$

And thus, we get: $F(q) = \dfrac{(q^2 + 3q + 2)(q^2 - 4q + 1)}{q^2 - 4q + 3}$.

And we know: $q^2 - 4q + 1 = 0$.

So we get: $F(q) = 0$.

Examples C in Irrationals

Simplify each expression below:

0. $\sqrt{\frac{7}{6} - \sqrt{\frac{4}{3}}}$

1. $\dfrac{1}{\sqrt{9 + 4\sqrt{4 + \sqrt{12}}}}$

2. $\sqrt{(a+b)^2 - \sqrt{8(a^3 b + ab^3)}}$

3. $\sqrt{a - \sqrt{a^2 - b^2}}$ where $a > b > 0$.

Suggestions or Solutions
To the Examples C in Irrationals

3.0. **Simplify:** $\sqrt{\frac{7}{6} - \sqrt{\frac{4}{3}}}$.

Multiplying 1 by a square root of a number twice, we get the number.
That is to say that squaring a square root, we get what's inside the root sign.

And doing algebra, we often use this tool: $a^2 \pm 2ab + b^2 = (a \pm b)^2$, which is often called a *complete square*. So doing this problem, we are going to use the tool above.

Now, assuming $R = \sqrt{\frac{7}{6} - \sqrt{\frac{4}{3}}}$, and setting: $\frac{7}{6} - \sqrt{\frac{4}{3}} = (a - b)^2$, we get: $R = \sqrt{(a-b)^2}$.

Then, we get first: $R > 0$, simply because a square root is ≥ 0, and $a - b \neq 0$.
So we get: $R = a - b$ if $a - b > 0$, that is, if $a > b$.

And of course, if $a < b$, that is, if $a - b < 0$, we get: $R = -(a - b) = b - a$.
So finding a and b, we get R. How?

Getting back to the complete square, we have: $(a - b)^2 = a^2 - 2ab + b^2 = a^2 + b^2 - 2ab$.

And we have: $\frac{7}{6} - \sqrt{\frac{4}{3}} = (a - b)^2$.

So converting $\frac{7}{6} - \sqrt{\frac{4}{3}}$ into a complete square, we can get a and b. How?

Looking closely at $a^2 + b^2 - 2ab$ and $\frac{7}{6} - \sqrt{\frac{4}{3}}$, we can notice that if 7/6 is <u>a sum of squares of two numbers</u>, and $\sqrt{\frac{4}{3}}$ is <u>twice the product of the two numbers</u>, we can put $\frac{7}{6} - \sqrt{\frac{4}{3}}$ into a complete square taking a form of $(\sqrt{u} - \sqrt{v})^2$.

That is to say that it can be the case where we get: $\frac{7}{6} - \sqrt{\frac{4}{3}} = (\sqrt{u} - \sqrt{v})^2$.

And in fact, we can get: $\frac{7}{6} - \sqrt{\frac{4}{3}} = \frac{7}{6} - 2\sqrt{\frac{1}{3}}$.   How then, can we get u and v?

What we want to get is: $\frac{7}{6} - \sqrt{\frac{4}{3}} = (\sqrt{u} - \sqrt{v})^2$.

And expanding the right hand side, we get: $u + v - 2\sqrt{uv}$.

Then, we get: $\frac{7}{6} - 2\sqrt{\frac{1}{3}} = u + v - 2\sqrt{uv}$.   What then, can we get?

We can get: $u + v = \frac{7}{6}$, and $uv = \frac{1}{3}$.   How then, can we get u and v?

We can set up an equation as follows: $x^2 - (u + v)x + uv = 0$.
Then, the solution is: $x = u$ or v.   How come?

We have: $x^2 - (u + v)x + uv = 0 \Rightarrow (x - u)(x - v) = 0 \Rightarrow x = u$ or v.

We now have: $u + v = \frac{7}{6}$, and $uv = \frac{1}{3}$.

So solving the equation $x^2 - \frac{7}{6}x + \frac{1}{3} = 0$, we can get u and v.

And thus, solving the equation, we get:

$x^2 - \frac{7}{6}x + \frac{1}{3} = 0 \Rightarrow 6x^2 - 7x + 2 = 0 \Rightarrow (3x - 2)(2x - 1) = 0 \Rightarrow x = \frac{2}{3}$ or $\frac{1}{2}$.

That is to say that $(x - \frac{2}{3})(x - \frac{1}{2}) = 0$.

And the equation we began with is: $(x - u)(x - v) = 0$.

So we can see that $(u, v) = (\frac{2}{3}, \frac{1}{2})$ or $(\frac{1}{2}, \frac{2}{3})$. Which one of the two then?

We have: $\frac{7}{6} - \sqrt{\frac{4}{3}} = (\sqrt{u} - \sqrt{v})^2$. And we have: $R = \sqrt{\frac{7}{6} - \sqrt{\frac{4}{3}}}$.

So we get: $R = \sqrt{(\sqrt{u} - \sqrt{v})^2}$. And we know: $R > 0$.

So if $\sqrt{u} > \sqrt{v}$, we get: $R = \sqrt{u} - \sqrt{v}$, and if $\sqrt{u} < \sqrt{v}$, we get: $R = \sqrt{v} - \sqrt{u}$.

And thus, either way, we want to get: $R > 0$.

So if we set: $R = \sqrt{(\sqrt{\frac{2}{3}} - \sqrt{\frac{1}{2}})^2}$, we get: $R = \sqrt{\frac{2}{3}} - \sqrt{\frac{1}{2}}$, since $2/3 > 1/2$.

And even if we set: $R = \sqrt{(\sqrt{\frac{1}{2}} - \sqrt{\frac{2}{3}})^2}$, we get: $R = \sqrt{\frac{2}{3}} - \sqrt{\frac{1}{2}}$, too.

And thus, we get: $\sqrt{\frac{7}{6} - \sqrt{\frac{4}{3}}} = \sqrt{\frac{2}{3}} - \sqrt{\frac{1}{2}}$.

3.1. Simplify: $\dfrac{1}{\sqrt{9 + 4\sqrt{4 + \sqrt{12}}}}$

Adding together the sum of squares of two numbers and twice the product of the two numbers, we get the square of the sum of the two numbers, and the square is called a complete square. That is to say that:

We have: $a^2 \pm 2ab + b^2 = a^2 + b^2 \pm 2ab = (a \pm b)^2$, often called a complete square.

So setting: $Q = \dfrac{1}{\sqrt{9 + 4\sqrt{4 + \sqrt{12}}}}$, we can simplify Q using the idea above.

First, we can convert $4+\sqrt{12}$ into a form of $(\sqrt{u}-\sqrt{v})^2$ the way below:

$$4+\sqrt{12} = 4+2\sqrt{3} = 3+1+2\sqrt{3\cdot 1} = (\sqrt{3}+\sqrt{1})^2 = (\sqrt{3}+1)^2.$$

Then, we get: $Q = \dfrac{1}{\sqrt{9+4\sqrt{4+\sqrt{12}}}} = \dfrac{1}{\sqrt{9+4\sqrt{(\sqrt{3}+1)^2}}} = \dfrac{1}{\sqrt{9+4(\sqrt{3}+1)}} = \dfrac{1}{\sqrt{13+4\sqrt{3}}}.$

So next, looking at $\sqrt{13+4\sqrt{3}}$, we can notice that we can get:

$$13+4\sqrt{3} = 13+2\sqrt{12} = 12+1+2\sqrt{12\cdot 1} = (\sqrt{12}+1)^2.$$

Thus, we get: $Q = \dfrac{1}{\sqrt{(\sqrt{12}+1)^2}} = \dfrac{1}{\sqrt{12}+1}.$

And making the denominator rational, that is, rationalizing the denominator, we get:

$$Q = = \dfrac{1}{\sqrt{12}+1} = \dfrac{\sqrt{12}-1}{(\sqrt{12}+1)(\sqrt{12}-1)} = \dfrac{\sqrt{12}-1}{12-1} = \dfrac{\sqrt{12}-1}{11}.$$

3.2. Simplify: $\sqrt{(a+b)^2 - \sqrt{8(a^3b+ab^3)}}$

We have: $a^2 \pm 2ab + b^2 = a^2 + b^2 \pm 2ab = (a \pm b)^2.$
So we have: $(a+b)^2 = a^2 + b^2 + 2ab.$
And we can get notice that: $a^3b + ab^3 = ab(a^2 + b^2).$
So we can get:

$$(a+b)^2 - \sqrt{8(a^3b+ab^3)} = a^2 + b^2 + 2ab - 2\sqrt{2ab(a^2+b^2)} = (\sqrt{a^2+b^2} - \sqrt{2ab})^2.$$

Thus, we get: $\sqrt{(a+b)^2 - \sqrt{8(a^3b+ab^3)}} = \sqrt{(\sqrt{a^2+b^2} - \sqrt{2ab})^2}$.

And we know: $a^2 + b^2 + 2ab = (a+b)^2 \geq 0 \Rightarrow a^2 + b^2 \geq 2ab$.

So we get: $\sqrt{(a+b)^2 - \sqrt{8(a^3b+ab^3)}} = \sqrt{a^2+b^2} - \sqrt{2ab}$.

3.3.   Simplify: $\sqrt{a - \sqrt{a^2-b^2}}$ where $a > b > 0$.

We have: $a - \sqrt{a^2-b^2} = \frac{2a-2\sqrt{a^2-b^2}}{2}$.

And we can get: $a^2 - b^2 = (a+b)(a-b)$.   So what?

We can get this, too: $2a = a + b + a - b = (a+b) + (a-b)$.

So we can get: $2a - 2\sqrt{a^2-b^2} = (a+b) + (a-b) - 2\sqrt{(a+b)(a-b)} = (\sqrt{a+b} - \sqrt{a-b})^2$.

And we have: $a > b > 0$.   So we get: $a + b > a - b$.

Thus, we get: $\sqrt{a - \sqrt{a^2-b^2}} = \sqrt{\frac{2a-2\sqrt{a^2-b^2}}{2}} = \sqrt{\frac{(\sqrt{a+b} - \sqrt{a-b})^2}{2}} = \frac{\sqrt{a+b} - \sqrt{a-b}}{\sqrt{2}}$.

Examples D in Irrationals

0. Assuming $a = \frac{1}{\sqrt{2}}$, find the value of $\sqrt{\frac{1+a}{1-a}} - \sqrt{\frac{1-a}{1+a}}$.

1. Assuming $b = \sqrt{3}$, find the value of $\dfrac{1}{\sqrt{b+1-2\sqrt{b}}} - \dfrac{1}{\sqrt{b+1+2\sqrt{b}}}$.

2. Assuming $c = \frac{\sqrt{3}}{4}$, find the value of $\dfrac{1+2c}{1+\sqrt{1+2c}} + \dfrac{1-2c}{1-\sqrt{1-2c}}$.

3. Assuming $d = \sqrt{2} - 1$, find the value of $\sqrt{\dfrac{d+1+\sqrt{d^2+2d}}{d+1-\sqrt{d^2+2d}}}$.

4. Simplify: $\sqrt[3]{10+\sqrt{108}} + \sqrt[3]{10-\sqrt{108}}$.

5. Find u and v for which $\dfrac{u}{3-2\sqrt{2}} - \dfrac{v}{3+2\sqrt{2}} = 3+6\sqrt{2}$.

6. Assuming $(u\sqrt{2}+\sqrt{3})u+(v\sqrt{2}+\sqrt{3})v+(w\sqrt{2}+\sqrt{3})w = 5\sqrt{2}+3\sqrt{3}$, find the value of $uv + vw + wu$.

Suggestions or Solutions
To the Examples D in Irrationals

0.   Assuming $a = \frac{1}{\sqrt{2}}$, find the value of $\sqrt{\frac{1+a}{1-a}} - \sqrt{\frac{1-a}{1+a}}$.

$$\sqrt{\frac{1+a}{1-a}} - \sqrt{\frac{1-a}{1+a}} = \frac{\sqrt{1+a}}{\sqrt{1-a}} - \frac{\sqrt{1-a}}{\sqrt{1+a}} = \frac{(\sqrt{1+a})^2 - (\sqrt{1-a})^2}{\sqrt{1-a}\sqrt{1+a}} = \frac{(1+a)-(1-a)}{\sqrt{(1-a)(1+a)}} = \frac{2a}{\sqrt{1-a^2}}.$$

And we have: $a = \frac{1}{\sqrt{2}}$.

So we get: $\dfrac{2a}{\sqrt{1-a^2}} = \dfrac{\frac{2}{\sqrt{2}}}{\sqrt{1-\frac{1}{2}}} = \dfrac{\frac{(\sqrt{2})^2}{\sqrt{2}}}{\sqrt{\frac{1}{2}}} = \dfrac{\sqrt{2}}{\frac{1}{\sqrt{2}}} = \dfrac{\sqrt{2}\cdot\sqrt{2}}{\frac{1}{\sqrt{2}}\cdot\sqrt{2}} = \dfrac{2}{1} = 2.$

1.   Assuming $b = \sqrt{3}$, find the value of $\dfrac{1}{\sqrt{b+1-2\sqrt{b}}} - \dfrac{1}{\sqrt{b+1+2\sqrt{b}}}$.

First, we can get:

$$b+1-2\sqrt{b} = b+1-2\sqrt{b\cdot 1} = (\sqrt{b}-1)^2, \text{ and } b+1+2\sqrt{b} = b+1+2\sqrt{b\cdot 1} = (\sqrt{b}+1)^2.$$

So we get: $\dfrac{1}{\sqrt{b+1-2\sqrt{b}}} - \dfrac{1}{\sqrt{b+1+2\sqrt{b}}} = \dfrac{1}{\sqrt{(\sqrt{b}-1)^2}} - \dfrac{1}{\sqrt{(\sqrt{b}+1)^2}} = \dfrac{1}{\sqrt{b}-1} - \dfrac{1}{\sqrt{b}+1}.$

Next, we can get: $\dfrac{1}{\sqrt{b}-1} - \dfrac{1}{\sqrt{b}+1} = \dfrac{(\sqrt{b}+1)-(\sqrt{b}-1)}{(\sqrt{b}-1)(\sqrt{b}+1)} = \dfrac{2}{b-1}.$ And we have: $b = \sqrt{3}$.

So we get: $\dfrac{1}{\sqrt{b+1-2\sqrt{b}}} - \dfrac{1}{\sqrt{b+1+2\sqrt{b}}} = \dfrac{2}{b-1} = \dfrac{2}{\sqrt{3}-1} = \dfrac{2(\sqrt{3}+1)}{3-1} = \sqrt{3}+1.$

2. Assuming $c = \frac{\sqrt{3}}{4}$, find the value of $\frac{1+2c}{1+\sqrt{1+2c}} + \frac{1-2c}{1-\sqrt{1-2c}}$.

First, we can get: $c = \frac{\sqrt{3}}{4} \Rightarrow 1 + 2c = 1 + \frac{2\sqrt{3}}{4} = \frac{4+2\sqrt{3}}{4} = \frac{3+1+2\sqrt{3\cdot1}}{4} = \frac{(\sqrt{3}+1)^2}{4}$, and also, we get:

$c = \frac{\sqrt{3}}{4} \Rightarrow 1 - 2c = 1 - \frac{2\sqrt{3}}{4} = \frac{4-2\sqrt{3}}{4} = \frac{3+1-2\sqrt{3\cdot1}}{4} = \frac{(\sqrt{3}-1)^2}{4}$.

So next, we get: $\sqrt{1+2c} = \sqrt{\frac{(\sqrt{3}+1)^2}{4}} = \frac{\sqrt{3}+1}{2}$, and $\sqrt{1-2c} = \sqrt{\frac{(\sqrt{3}-1)^2}{4}} = \frac{\sqrt{3}-1}{2}$.

Thus next, we get: $1 + \sqrt{1+2c} = 1 + \frac{\sqrt{3}+1}{2} = \frac{3+\sqrt{3}}{2}$, and $1 - \sqrt{1-2c} = 1 - \frac{\sqrt{3}-1}{2} = \frac{3-\sqrt{3}}{2}$.

And we have: $1 + 2c = \frac{4+2\sqrt{3}}{4} = \frac{2+\sqrt{3}}{2}$, and $1 - 2c = \frac{4-2\sqrt{3}}{4} = \frac{2-\sqrt{3}}{2}$.

So we get: $\frac{1+2c}{1+\sqrt{1+2c}} = \frac{\frac{2+\sqrt{3}}{2}}{\frac{3+\sqrt{3}}{2}} = \frac{2+\sqrt{3}}{3+\sqrt{3}}$, and $\frac{1-2c}{1-\sqrt{1-2c}} = \frac{\frac{2-\sqrt{3}}{2}}{\frac{3-\sqrt{3}}{2}} = \frac{2-\sqrt{3}}{3-\sqrt{3}}$.

Thus, we get: $\frac{1+2c}{1+\sqrt{1+2c}} + \frac{1-2c}{1-\sqrt{1-2c}} = \frac{2+\sqrt{3}}{3+\sqrt{3}} + \frac{2-\sqrt{3}}{3-\sqrt{3}} = \frac{(2+\sqrt{3})(3-\sqrt{3})+(2-\sqrt{3})(3+\sqrt{3})}{9-3}$

$= \frac{(6+\sqrt{3}-3)+(6-\sqrt{3}-3)}{6} = \frac{6}{6} = 1$.

3. Assuming $d = \sqrt{2} - 1$, find the value of $\sqrt{\frac{d+1+\sqrt{d^2+2d}}{d+1-\sqrt{d^2+2d}}}$.

First, we get: $d = \sqrt{2} - 1 \Rightarrow d + 1 = \sqrt{2}$.

So next, we get: $(d+1)^2 = 2 \Rightarrow d^2 + 2d + 1 = 2 \Rightarrow d^2 + 2d = 1$.

And thus, we get: $\sqrt{\frac{d+1+\sqrt{d^2+2d}}{d+1-\sqrt{d^2+2d}}} = \sqrt{\frac{\sqrt{2}+1}{\sqrt{2}-1}} = \sqrt{\frac{(\sqrt{2}+1)^2}{2-1}} = \sqrt{2} + 1$.

4. Simplify: $\sqrt[3]{10+\sqrt{108}} + \sqrt[3]{10-\sqrt{108}}$.

We have: $a^3 + b^3 = (a+b)^3 - 3ab(a+b)$.

So assuming first, $a = \sqrt[3]{10+\sqrt{108}}$, and $b = \sqrt[3]{10-\sqrt{108}}$, we want to find $a+b$.

Then first, we get: $a^3 = 10+\sqrt{108}$, and $b^3 = 10-\sqrt{108}$, so we get: $a^3 + b^3 = 20$.
And next, we get:

$$ab = \sqrt[3]{10+\sqrt{108}} \cdot \sqrt[3]{10-\sqrt{108}} = \sqrt[3]{(10+\sqrt{108})(10-\sqrt{108})} = \sqrt[3]{100-108} = \sqrt[3]{-8} = -2.$$

So we get: $a^3 + b^3 = (a+b)^3 - 3ab(a+b) \Rightarrow 20 = (a+b)^3 + 6(a+b)$.
How then, can we find $a+b$?

Setting: $a + b = c$, we get: $20 = c^3 + 6c \Rightarrow c^3 + 6c - 20 = 0$.
So we want to solve the equation for c. How?

Factorizing it, we can get c.
Assuming $c - 2$ is a factor, we can do a synthetic division the way below:

```
2 | 1   0   6   -20
  |     2   4    20
  ———————————————————
    1   2  10    0
```

So we get: $(c-2)(c^2 + 2c + 10) = 0$.
And we get: $c^2 + 2c + 10 = c^2 + 2c + 1 + 9 = (c+1)^2 + 9 > 0$.

So we get: $c = 2$. And we know: $a + b = c$, where $a + b = \sqrt[3]{10+\sqrt{108}} + \sqrt[3]{10-\sqrt{108}}$.

So we get: $\sqrt[3]{10+\sqrt{108}} + \sqrt[3]{10-\sqrt{108}} = 2$.

5. **Find u and v for which** $\dfrac{u}{3-2\sqrt{2}} - \dfrac{v}{3+2\sqrt{2}} = 3+6\sqrt{2}.$

First, simplifying the left hand side, we get:

$$\frac{u}{3-2\sqrt{2}} - \frac{v}{3+2\sqrt{2}} = \frac{u(3+2\sqrt{2})-v(3-2\sqrt{2})}{9-8} = 3u-3v+(u+v)2\sqrt{2}.$$

So we get: $\dfrac{u}{3-2\sqrt{2}} - \dfrac{v}{3+2\sqrt{2}} = 3+6\sqrt{2} \Rightarrow 3u-3v+(u+v)2\sqrt{2} = 3+6\sqrt{2}$

$\Rightarrow 3u - 3v - 3 + (2u + 2v - 6)\sqrt{2} = 0.$

Thus, we get: $3u - 3v - 3 = 0$, and $2u + 2v - 6 = 0$.

And next, solving the system above, we get first:

$3u - 3v - 3 = 0 \Rightarrow u - v = 1$, and $2u + 2v - 6 = 0 \Rightarrow u + v = 3$.

So next, we get: $(u - v) + (u + v) = 1 + 3 = 4 \Rightarrow 2u = 4 \Rightarrow u = 2$.

And next, we have: $u - v = 1$. So we get: $2 - v = 1 \Rightarrow v = 1$.

6. Assuming $(u\sqrt{2}+\sqrt{3})u+(v\sqrt{2}+\sqrt{3})v+(w\sqrt{2}+\sqrt{3})w = 5\sqrt{2}+3\sqrt{3}$, find the value of $uv + vw + wu$.

First, rewriting the left hand side, we get:

$$(u\sqrt{2}+\sqrt{3})u+(v\sqrt{2}+\sqrt{3})v+(w\sqrt{2}+\sqrt{3})w = (u^2+v^2+w^2)\sqrt{2}+(u+v+w)\sqrt{3}.$$

So we get: $(u\sqrt{2}+\sqrt{3})u+(v\sqrt{2}+\sqrt{3})v+(w\sqrt{2}+\sqrt{3})w = 5\sqrt{2}+3\sqrt{3}$

$\Rightarrow (u^2+v^2+w^2-5)\sqrt{2}+(u+v+w-3)\sqrt{3} = 0.$ So we get:

$u^2+v^2+w^2-5 = 0 \Rightarrow u^2+v^2+w^2 = 5$, and $u+v+w-3 = 0 \Rightarrow u+v+w = 3$.

And we have: $(u+v+w)^2 = u^2+v^2+w^2+2(uv+vw+wu)$.

So we get: $3^2 = 5 + 2(uv+vw+wu) \Rightarrow uv+vw+wu = 2$.

Sense of Arithmetic 5

Read the expressions below filling in the blanks by calculation by heart. If however, you want to put down the solutions, use a separate sheet of paper. And just keep reading them at your pace. And of course, you don't have to read them all at once.

Each reading can take 5, 10, or 15 minutes at a time, or as much as you can concentrate. It can help increase sense of arithmetic.

$0 + \_ = 0 = \_ \times 2$

$0 - 0 = \_$

$\_ + 1 = 1$

$\_ - 1 = -1$

$1 - \_ = 1$

$1 + \_ = 2 = \_ \times 2$

$1 - \_ = 0$

$2 + 1 = \_ = 1 \times \_ + 1 = 1 + 1 + \_ = 1 \times \_ = 3$

$2 - 1 = 1 + \_ - 1 = 1 + \_ = 1$

$1 - \_ = 1 - (1 + 1) = 1 - 1 - \_ = 0 - 1 = -1$

$3 + \_ = 5 = 1 \times 3 + 1 \times \_ = 1 \times (3 + \_) = 1 \times \_ = 5$

$3 - 2 = 1 + \_ - 2 = 1 + \_ = 1$

$2 - \_ = 2 - (2 + 1) = 2 - \_ - 1 = \_ - 1 = -1$

$5 + \_ = 8, 5 - \_ = 2,$ and $3 - \_ = -2$

$8 + \_$
$= 8 + (2 + 2) = (8 + \_) + 2 = 10 + \_ = 12$
$= (2 + \_) + 4 = 2 + (\_ + 4) = 2 + 10 = 12$

$8 - \_ = (4 + \_) - 4 = 4 + (\_ - 4) = 4 + 0 = 4$

$10 + \_$
$= 10 \times (1 + \_) + 1 \times (0 + \_)$
$= 10 \times \_$
$= 10$

$10 - 0 = 10 = 10 \times (1 - \_) + 1 \times (0 - \_) = 10 + \_ = 10$

$10 + \_ = 11$

$10 - 1 = (0 + 1 \times \_) - 1 = 0 + (1 \times \_ - 1)$
$= 0 + (\_ - 1) = 0 + (9 + \_ - 1) = 0 + \_$

$1 - 10 = 1 - (0 + 1 \times \_) = 1 - (0 + 1 + 1 \times \_) = 1 - 1 - (0 + 1 \times \_)$
$= 0 - (0 + \_) = 0 - \_$

$10 + \_ = 19$

$10 - 9 = (0 + 1 \text{ x } \_) - 9 = (0 + (1 + \_)) - 9 = 1 + \_ - 9 = \_ + 0$

$10 + \_ = 20 = 10 \text{ x } \_ = 20$

$10 - 10 = 10 \text{ x } (1 - \_) = 10 \text{ x } \_ = \_$

$10 + \_\_ = 21$

$10 - 11 = 10 - (\_\_ + 1) = 10 - 10 + \_ = 0 + \_ = -1$

$11 - 10 = (10 + \_) - (10 + \_) = 10 \text{ x } (1 - \_) + 1 \text{ x } (1 - \_) = 0 + \_ = \_$

$10 + \_ = 29$

$10 - 19$
$= 10 - (10 + \_)$
$= (10 - \_) + (0 - \_) = 0 + \_ = -9$

$19 - 10$
$= (10 + \_) - 10$
$= (10 - \_) + (\_ - 0)$
$= 0 + \_$
$= 9$

$14 + \_ = 18$

$14 - 4 = \_ \times (1 - 0) + 1 \times (4 - 4) = 10 \times 1 + 1 \times \_ = 10 + \_ = 10$

$4 - 14 = 10 \times (0 - \_) + 1 \times (4 - 4) = 10 \times (-1) + 1 \times \_ = 10 \times (-1) + \_ = \_$

$16 + 9 = 16 + (\_ + 5) = (16 + \_) + 5 = ((10 + \_) + 4) + 5 = (10 + (\_ + 4)) + 5$
$= (10 + \_) + 5 = 10 \times \_ + 5 = 20 + \_$

$16 + 9 = (10 + \_) + 9 = (10 + (5 + \_)) + 9 = ((10 + 5) + \_) + 9$
$= (10 + \_) + (\_ + 9)$

$= (10 + \_) + 10 = 10 + \_ + 10 = 10 + \_ + 5 = 10 \times \_ + 5 = 20 + 5 = 25$

$16 - 9 = 10 \times (\_ - 0) + 1 \times (6 - \_) = 10 \times 1 + \_ = 10 + \_ = 7$

$9 - 16 = 10 \times (0 - \_) + 1 \times (9 - \_) = 10 \times \_ + 1 \times \_ = -10 + 3$
$= 3 + \_ = 3 - \_ = 1 \times (3 - \_) = 1 \times \_ = -7$

$-10 + 3 = - (7 + \_) + 3 = 3 + (- (7 + \_)) = 3 + (- (\_ + 7)) = 3 - (\_ + 7)$
$= \_ - 3 - 7 = 0 - 7 = -7$

39 + 58 = (30 + _) + (_ + 8) = 30 + _ + 9 + _ = 30 + _ + 7 + (_ + 8)

= 30 + _ + 7 + _ = _ + 7 = _

30 + _ + 7 + _

= 10 x (3 + _ + 1) + 7

= 10 x (_ + 1) + 7

= 10 x _ + 7

= 90 + _

= _

39 − 58

= 10 x (_ − 5) + 1 x (_ − 8)

= 10 x _ + 1 x _

= 10 x _ + ((-1) x _ + 1)

= -10 + (1 + _ x 10)

= -10 + (1 + _ x (1 + 9))

= -10 + (1 + ((-1) + _))

= -10 + (1 + _) + (-9)

= -10 + (_ − 1) + (-9)

= -10 + _ + (-9)

= -10 + _

= (-10) x _ + (-1) x _ = _

$78 + 85 + 69$

$= (70 + 80 + 60) + (8 + 5 + 9)$

$= (70 + 80 + 60) + (8 + (2 + \_) + 9)$

$= (70 + 80 + 60) + ((\_ + 3) + 9)$

$= (70 + 80 + 60) + (\_ + (3 + 9))$

$= (70 + 80 + 60) + (\_ + ((2 + 1) + 9))$

$= (70 + 80 + 60) + (\_ + (2 + 10))$

$= (70 + 80 + 60) + (\_ \times 2 + 2)$

$= 10 \times (7 + 8 + 6 + 2) + 1 \times \_$

$= 10 \times (15 + 6 + 2) + 1 \times \_$

$= 10 \times (21 + 2) + \_ \times 2$

$= 10 \times 23 + 1 \times \_$

$= 10 \times (10 \times \_ + 3) + 1 \times 2$

$= 100 \times \_ + 10 \times \_ + 1 \times 2$

$= 200 + 30 + 2 = 232$

$98 - 24 - 58$

$= (98 - 24) - 58$

$= 74 - 58$

$= 16$

$(98 - 24) - 58$

$= 98 - (\_ + 58)$

$= 98 - \_$

$= 16$

$34 - 45 - 76 + 27$

$= 34 - (45 + \_) + 27$

$= 34 + 27 - (45 + \_)$

$= 61 - \_$

$= (60 + 1) + ((-100) + (-20) + (-1))$

$= -100 + (\_ + (-20)) + (1 + (-1))$

$= -100 + \_ + 0$

$= -100 \times 1 + \_ \times 4 + 1 \times 0$

$= -100 \times 0 + (\_ \times 4 + (-10) \times 10) + 0$

$= 0 + 10 \times 4 + 10 \times \_ + 0$

$= 10 \times (4 + \_)$

$= 10 \times \_$

$= \_ \times 10$

$= -60$

$= 34 + 27 + \_ - 76$

$= 61 + (-(\_ + 76))$

$= 61 + \_$

$= 61 - (\_ + 60)$

$= 61 + \_ - 60$

$= \_$

$0 \times 0 = \_, \quad 0 \times 1 = \_, \quad 0 \times (-1) = \_, \quad 1 \times 0 = \_, \quad$ and $(-1) \times 0 = \_$

$3 \times 2 = 3 + \_ = 6$, and $(-3) \times \_ = -3 + (-3) = -6$

3 x (-2) = 3 x (-1) x _ = _ x 3 x 2 = -1 x _ = _

-3 x (-2) = -1 x 3 x _ x 2 = -1 x _ x 3 x 2 = _

12 x _ = 12 + 12 + 12 = 36 = (10 + _) x 3 = 10 x 3 + _ x 3 = 30 + _ = _

15 x 12

= (10 + _) x 12

= 10 x 12 + _ x 12

= 10 x (10 + _) + 5 x (10 + _)

= 10 x _ + 10 x 2 + 5 x _ + 5 x 2

= _ + 20 + 50 + 10

= _

0.1 + _ = 0.2   0.1 − _ = 0

4.4 + 32.7

= (1 x 4 + 0.1 x _) + (10 x _ + 1 x 2 + 0.1 x 7)

= 10 x (0 + _) + 1 x (4 + _) + 0.1 x (_ + 7)

= 10 x _ + 1 x (4 + _) + 0.1 x (_ + 1)

= 10 x _ + 1 x (4 + _ + 1) + 0.1 x 1

= 30 + _ + 0.1

= _

4.4 − 32.7

$= (1 \times \_ + 0.1 \times 4) - (10 \times \_ + 1 \times 2 + 0.1 \times \_)$

$= 10 \times (0 + \_) + 1 \times (4 + (-2)) + 0.1 \times (\_ + (-7))$

$= \_ \times (-3) + 1 \times 2 + 0.1 \times (-3)$

$= 10 \times ((-2) + (-1)) + 1 \times \_ + 0.1 \times (-3)$

$= 10 \times (-2) + \_ \times (-1) + 1 \times \_ + 0.1 \times (-3)$

$= 10 \times (-2) + (-10) + 1 \times \_ + 0.1 \times (-3)$

$= 10 \times (-2) + 1 \times \_ + 1 \times \_ + 0.1 \times (-3)$

$= 10 \times (-2) + 1 \times (\_ + 2) + 0.1 \times (-3)$

$= 10 \times \_ + 1 \times (-8) + 0.1 \times (-3)$

$= \_$

1.8 × _

$= 1.8 + 1.8$

$= 1 \times (1 + \_) + 0.1 \times (\_ + 8)$

$= 1 \times (1 + \_) + 0.1 \times (\_ + 6)$

$= 1 \times (1 + \_ + 1) + 0.1 \times \_$

$= 1 \times \_ + 0.1 \times \_$

$= \_$

1.8 + 1.8

= (1 + 0.8) x _ = 1 x 2 + 0.8 x 2

= 1 x 2 + 0.1 x _ x 2

= 1 x 2 + 0.1 x _

= 1 x 2 + 0.1 x (_ + 6)

= 1 x2 + 0.1 x 10 + 0.1 x _

= 1 x (_ + 1) + 0.1 x 6

= 1 x _ + 0.1 x _

= _

0.3 x 0.2 = 0.1 x 3 x 0.1 x _ = 0.1 x 0.1 x 3 x _ = 0.01 x _ = _

0.3 x 0.8

= 0.1 x _ x 0.1 x 8

= 0.1 x 0.1 x _ x 8

= 0.01 x _

= 0.01 x (_ + 4)

= 0.01 x 20 + 0.01 x _

= 0.01 x _ x 10 + 0.01 x _

= 0.01 x 10 x _ + 0.01 x 4

= 0.1 x _ + 0.01 x 4

= _ + 0.04

= _

1.9 x 0.7

= (1 + 0.9) x 0.7

= 1 x _ + 0.9 x 0.7

= 0.7 + 0.1 x _ x 9 x 7

= 0.7 + 0.01 x (60 + _)

= 0.7 + 0.01 x (10 x _ + 3)

= 0.7 + (0.01 x 10) x _ + 0.03

= 0.7 + 0.1 x _ + 0.03

= 0.7 + _ + 0.03

= 0.1 x (_ + 6) + 0.03

= 0.1 x (_ + 3) + 0.03

= 0.1 x 10 + _ + 0.03

= _ + 0.3 + 0.03

= _

5.9 x 9.8

= (5 + _) x (_ + 0.8)

= 5 x (_ + 0.8) + 0.9 x (_ + 0.8)

= 5 x 9 + _ x 0.8 + 0.9 x 9 + 0.9 x 0.8

= 45 + 5 x (0.1 x _) + 0.1 x (9 x _) + 0.1 x 0.1 x (9 x _)

= 45 + 5 x (8 x 0.1) + 0.1 x 81 + 0.01 x _

= 45 + (5 x _) x 0.1 + 0.1 x 81 + 0.01 x _

= 45 + 40 x 0.1 + 0.1 x 81 + 0.01 x _

= 45 + 0.1 x 40 + 0.1 x 81 + 0.01 x _

$= 45 + 0.1 \times (40 + \_) + 0.01 \times \_ = 45 + 0.1 \times 121 + 0.01 \times 72$

$= 40 + 5 + 0.1 \times (100 + \_ + 1) + 0.01 \times (70 + 2)$

$= 40 + 5 + 0.1 \times 100 + 0.1 \times \_ + 0.1 + 0.01 \times \_ + 0.01 \times 2$

$= 40 + 5 + 0.1 \times 10 \times \_ + 0.1 \times 10 \times 2 + 0.1 + 0.01 \times 10 \times \_ + 0.01 \times 2$

$= 40 + 5 + \_ + 2 + 0.1 + 0.1 \times 7 + 0.01 \times 2$

$= 50 + 7 + 0.1 \times (1 + \_) + 0.01 \times 2 = 50 + 7 + 0.1 \times \_ + 0.01 \times 2 = \_$

0.04×7.9

$= 0.04 \times (\_ + 0.9)$

$= 0.04 \times 7 + \_ \times 0.9$

$= (0.01 \times \_) \times 7 + (0.01 \times \_) \times 0.9$

$= 0.01 \times (\_ \times 7) + (0.01 \times 4) \times (\_ \times 9)$

$= 0.01 \times (\_ \times 7) + 0.01 \times \_ \times 4 \times 9$

$= 0.01 \times \_ + 0.001 \times \_$

$= 0.01 \times \_ + 0.001 \times (30 + \_)$

$= 0.01 \times \_ + 0.001 \times (10 \times \_ + \_)$

$= 0.01 \times \_ + 0.01 \times \_ + 0.001 \times \_$

$= 0.01 \times (28 + \_) + 0.001 \times \_$

$= 0.01 \times \_ + 0.006$

$= 0.01 \times (30 + \_) + 0.006$

$= 0.01 \times (10 \times \_) + 0.01 \times \_ + 0.006$

$= \_ \times 3 + 0.01 \times \_ + 0.006$

$= \_ + 0.01 + 0.006$

$= \_$

1 / _ = 1 since 1 = 1 x 1, and -1 / 1 = _ since -1 = 1 x (-1) = (-1) x 1

1 / (-1) = _ since 1 = (-1) x (-1) = 1

-1 / (-1) = _ since -1 = (-1) x 1 = -1, and 2 / 1 = 2 since 2 = 1 x 2

-2 / 1 = _ since -2 = 1 x (-2) = (-2) x 1

2 / (-1) = _ since 2 = (-1) x (-2) = (-1) x (_ x 2) = ((-1) x _) x 2 = 1 x 2 = 2

2 / 2 = 1 since 2 = 2 x 1, and -2 / 2 = _ since -2 = 2 x (-1) = (-1) x 2 = -2

2 / -2 = _ since 2 = (-2) x (-1) = (-1) x 2 x (-1) = (-1) x (-1) x 2 = 1 x 2

4 / 2 = _ since 4 = 2 x _, and 6 / _ = 2 since 6 = _ x 2

10 / _ = 10 since 10 = 1 x _, and 12 / _ = 4 since 12 = _ x 4

28 / 2 = (20 + _) / 2 = (20 / 2) + (_ / 2) = 10 + _ = _

= (_ − 2) / 2 = (_ / 2) − (2 / 2) = _ − 1 = _

1248 / 4

$= (1200 + 40 + 8) / 4$

$= (1200 / \_) + (40 / \_) + (8 / \_)$

$= ((100 \text{ x } \_) / 4) + ((10 \text{ x } \_) / 4) + ((1 \text{ x } \_) / 4)$

$= (100 \text{ x } (\_ / 4)) + (10 \text{ x } (\_ / 4)) + (1 \text{ x } (\_ / 4))$

$= 100 \text{ x } \_ + 10 \text{ x } \_ + 1 \text{ x } \_$

$= 300 + 10 + 2$

$= 312$

$1248 / 4 / 2 = (1248 / \_) / 2 = 1248 / (\_ \text{ x } 2) = 1248 / \_ = 156$

1248 / 4 / 2

$= 312 / \_$

$= (100 \text{ x } \_ + 10 \text{ x } \_ + 1 \text{ x } \_) / \_$

$= 100 \text{ x } (3 / 2) + 10 \text{ x } (1 / 2) + 1 \text{ x } (2 / 2)$

$= 100 \text{ x } (\_ + 1 / 2) + 10 \text{ x } (1 / \_) + 1 \text{ x } (2 / \_)$

$= 100 \text{ x } 1 + 100 / \_ + 10 / \_ + 1 \text{ x } \_$

$= 100 \text{ x } 1 + (10 \text{ x } \_) / 2 + (1 \text{ x } \_) / 2 + 1 \text{ x } 1$

$= 100 \text{ x } 1 + 10 \text{ x } (\_ / 2) + 1 \text{ x } (\_ / 2) + 1 \text{ x } 1$

$= 100 \text{ x } 1 + 10 \text{ x } \_ + 1 \text{ x } \_ + 1 \text{ x } 1$

$= 100 \text{ x } 1 + 10 \text{ x } \_ + 1 \text{ x } (\_ + 1)$

$= 100 \text{ x } 1 + 10 \text{ x } \_ + 1 \text{ x } \_$

$= \_$

1 / 2

= (0.1 x _) / 2 = 0.1 x (_ / 2) = 0.1 x _ · = _

= 0.5 since 0.5 x _ = 1

0.1 / 2

= (0.01 x _) / 2 = 0.01 x (_ / 2) = 0.01 x _ = _

= 0.05 since 0.05 x _ = 0.1

3 / 2 = (2 + _) / 2 = 2 / 2 + _ / 2 = 1 + _ = _

15 / 6

= (12 + _) / 6

= 12 / 6 + _ / 6

= 2 + (1 x _) / 6

= 2 + ((0.1 x _) x 3) / 6

= 2 + (0.1 x (_ x 3)) / 6

= 2 + (0.1 x _) / 6

= 2 + 0.1 x (_ / 6)

= 2 + 0.1 x _

= 2 + _

= _

3.9 / 2

= (3 + _) / 2

= (1 x 3 + 0.1 x _) / 2

= (1 x 3) / 2 + (0.1 x _) / 2

= 1 x (_ / 2) + 0.1 x (_ / 2)

= 1 x ((_ + 1) / 2) + 0.1 x ((_ + 1) / 2)

= 1 x ((2 / 2) + (1 / 2)) + 0.1 x ((_ / 2) + (1 / 2))

= 1 x (1 + _) + 0.1 x (_ + 0.5)

= 1 x 1 + 0.5 + 0.4 + 0.05

= 1 x 1 + 0.1 x (_ + 4) + 0.01 x _

= 1 x 1 + 0.1 x _ + 0.01 x _

= _

2 / 4 = (1 x _) / 4 = ((0.1 x _) x 2) / 4 = (0.1 x (_ x 2)) / 4 = (0.1 x _) / 4

= 0.1 x (_ / 4) = 0.1 x _ = _

1 / 4 = (0.1 x _) / 4 = 0.1 x (_ / 4) = 0.1 x ((_ + 2) / 4)

= 0.1 x ((_ / 4) + (2 / 4))

= 0.1 x (2 + _) = 0.1 x 2 + _ = 0.1 x 2 + _ x 5 = _

9.7 / 4

$= (9 + \_) / 4$

$= 9 / 4 + \_ / 4$

$= (8 + 1) / 4 + (0.1 \text{ x } \_) / 4$

$= 2 + 1 / 4 + 0.1 \text{ x } (\_ / 4)$

$= 2 + 0.25 + 0.1 \text{ x } ((\_ + 3) / 4)$

$= 2 + 0.25 + 0.1 \text{ x } (\_ + 3 / 4)$

$= 2 + 0.25 + 0.1 \text{ x } (1 + (0.1 \text{ x } \_) / 4)$

$= 2 + 0.25 + 0.1 \text{ x } (1 + 0.1 \text{ x } (\_ / 4))$

$= 2 + 0.25 + 0.1 \text{ x } (1 + 0.1 \text{ x } ((\_ + 2) / 4))$

$= 2 + 0.25 + 0.1 \text{ x } (1 + 0.1 \text{ x } (\_ + 2 / 4))$

$= 2 + 0.25 + 0.1 \text{ x } (1 + 0.1 \text{ x } (\_ + 0.5))$

$= 2 + 0.25 + 0.1 \text{ x } (1 + \_ + 0.05)$

$= 2 + 0.25 + 0.1 + \_ + 0.005$

$= 2 + 0.2 + \_ + 0.1 + 0.07 + 0.005$

$= 2 + 0.1 \text{ x } (\_ + 1) + 0.01 \text{ x } (5 + 7) + 0.005$

$= 2 + 0.1 \text{ x } \_ + 0.01 \text{ x } (10 + 2) + 0.005$

$= 2 + 0.1 \text{ x } \_ + 0.01 \text{ x } 10 + 0.01 \text{ x } 2 + 0.005$

$= 2 + 0.1 \text{ x } \_ + 0.1 \text{ x } 1 + 0.01 \text{ x } 2 + 0.005$

$= 2 + 0.1 \text{ x } \_ + 0.01 \text{ x } 2 + 0.005$

$= \_$

$1 / 0.1 = (0.1 \text{ x } \_) / 0.1 = (10 \text{ x } \_) / 0.1 = 10 \text{ x } (\_ / 0.1) = 10 \text{ x } 1$

1 / 0.2

= (0.1 x _) / 0.2

= (10 x _) / 0.2

= 10 x (_ / 0.2)

= 10 x ((0.01 x _) / 0.2)

= 10 x (0.01 x (_ / 0.2))

= 10 x (0.01 x ((0.1 x _) / 0.2))

= 10 x (0.01 x ((0.1 x (2 x _)) / 0.2))

= 10 x (0.01 x (((0.1 x 2) x _) / 0.2))

= 10 x (0.01 x ((0.2 x _) /0.2))

= 10 x (0.01 x ((_ x 0.2) / 0.2))

− 10 x (0.01 x (_ x (0.2 / 0.2)))

= 10 x (0.01 x (_ x 1))

= 10 x (0.01 x (10 x _))

= 10 x (0.1 x _)

= (10 x 0.1) x _

= 1 x _

= _

$2 / 0.4 = (0.1 \text{ x } \_) / 0.4 = (0.1 \text{ x } (4 \text{ x } \_)) / 0.4 = ((0.1 \text{ x } 4) \text{ x } \_) / 0.4$

$= (0.4 \text{ x } \_) / 0.4 = (\_ \text{ x } 0.4) / 0.4 = \_ \text{ x } (0.4 / 0.4) = \_ \text{ x } 1 = \_$

$7.9 / 0.4$

$= (7 + 0.9) / 0.4$

$= 7 / \_ + 0.9 / \_$

$= (4 + \_) / 0.4 + (0.1 \text{ x } \_) / 0.4$

$= 4 / 0.4 + 3 / 0.4 + 0.1 \text{ x } (9 / 0.4)$

$= 10 + (1 \text{ x } \_) / 0.4 + 0.1 \text{ x } (1 \text{ x } \_) / 0.4$

$= 10 + ((0.1 \text{ x } \_) \text{ x } 3) / 0.4 + 0.1 \text{ x } 0.1 \text{ x } \_ \text{ x } 9 / 0.4$

$= 10 + (0.1 \text{ x } \_) / 0.4 + 0.1 \text{ x } (0.1 \text{ x } \_) / 0.4$

$= 10 + 0.1 \text{ x } ((\_ + 2) / 0.4) + 0.1 \text{ x } 0.1 \text{ x } (\_ + 2) / 0.4$

$= 10 + 0.1 \text{ x } (\_ / 0.4 + 5) + 0.1 \text{ x } 0.1 \text{ x } (\_ / 0.4 + 5)$

$= 10 + 0.1 \text{ x } ((\_ \text{ x } 4) / 0.4 + 5) + 0.1 \text{ x } 0.1 \text{ x } (22 \text{ x } 4 / 0.4 + 5)$

$= 10 + 0.1 \text{ x } (\_ \text{ x } (4 / 0.4) + 5) + 0.1 \text{ x } 0.1 \text{ x } (22 \text{ x } (4 / 0.4) + 5)$

$= 10 + 0.1 \text{ x } (\_ \text{ x } 10 + 5) + 0.1 \text{ x } 0.1 \text{ x } (22 \text{ x } 10 + 5)$

$= 10 + 0.1 \text{ x } \_ \text{ x } 10 + 0.1 \text{ x } \_ + 0.1 \text{ x } (0.1 \text{ x } (22 \text{ x } 10) + 0.1 \text{ x } 5)$

$= 10 + 7 + \_ + 0.1 \text{ x } (22 + 0.5)$

$= 10 + 7 + \_ + 0.1 \text{ x } (20 + \_) + 0.05$

$= 10 + 7 + \_ + 0.1 \text{ x } (2 \text{ x } 10) + 0.2 + 0.05$

$= 10 + 7 + \_ + 2 + 0.2 + 0.05$

$= 10 + 9 + \_ + 0.05$

$= \_$

The numbers in bold below are the solutions, of course. Solutions alone however, don't mean much. So you may want to just keep reading the calculations below, if you want to, of course.

Note however, you don't have to do calculations the way below.

Each calculation just shows that **there are many ways we can calculate**.

$0 + \mathbf{0} = 0 = \mathbf{0}$ x 2, and $0 - 0 = \mathbf{0}$

$\mathbf{0} + 1 = 1$, and $\mathbf{0} - 1 = \text{-}1$

$1 - \mathbf{0} = 1$
$1 + \mathbf{1} = 2 = \mathbf{1}$ x 2
$1 - \mathbf{1} = 0$

$2 + 1 = \mathbf{3} = 1$ x $\mathbf{2} + 1 = 1 + 1 + \mathbf{1} = 1$ x $\mathbf{3} = 3$

$2 - 1 = 1 + \mathbf{1} - 1 = 1 + \mathbf{0} = 1$

$1 - \mathbf{2} = 1 - (1 + 1) = 1 - 1 - \mathbf{1} = 0 - 1 = \text{-}1$

$3 + \mathbf{2} = 5 = 1$ x $3 + 1$ x $\mathbf{2} = 1$ x $(3 + 2) = 1$ x $\mathbf{5} = 5$

$3 - \mathbf{2} = 1 + \mathbf{2} - 2 = 1 + \mathbf{0} = 1$

$2 - \mathbf{3} = 2 - (2 + 1) = 2 - \mathbf{2} - 1 = \mathbf{0} - 1 = \text{-}1$

$5 + 3 = 8$, $5 - \mathbf{3} = 2$, and $3 - \mathbf{5} = \text{-}2$

$8 + \mathbf{4}$

$= 8 + (2 + 2) = (8 + \mathbf{2}) + 2 = 10 + \mathbf{2} = 12$

$= (2 + \mathbf{6}) + 4 = 2 + (\mathbf{6} + 4) = 2 + 10 = 12$

$8 - \mathbf{4} = (4 + \mathbf{4}) - 4 = 4 + (\mathbf{4} - 4) = 4 + 0 = 4$

$10 + \mathbf{0} = 10 = 10 \times (1 + \mathbf{0}) + 1 \times (0 + \mathbf{0}) = 10 \times \mathbf{1} = 10$

$10 - 0 = 10 = 10 \times (1 - \mathbf{0}) + 1 \times (0 - \mathbf{0}) = 10 + \mathbf{0} = 10$

$10 + \mathbf{1} = 11$

$10 - 1 = (0 + 1 \times \mathbf{10}) - 1 = 0 + (1 \times \mathbf{10} - 1) = 0 + (\mathbf{10} - 1)$

$= 0 + (9 + \mathbf{1} - 1) = 0 + \mathbf{9}$

$1 - 10 = 1 - (0 + 1 \times \mathbf{10}) = 1 - (0 + 1 + 1 \times \mathbf{9}) = 1 - 1 - (0 + 1 \times \mathbf{9})$

$= 0 - (0 + \mathbf{9}) = 0 - \mathbf{9}$

$10 + \mathbf{9} = 19$

$10 - 9 = (0 + 1 \times \mathbf{10}) - 9 = (0 + (1 + \mathbf{9})) - 9 = 1 + \mathbf{9} - 9 = \mathbf{9} + 0$

$10 + \mathbf{10} = 20 = 10 \times \mathbf{2} = 20$

$10 - 10 = 10 \times (1 - \mathbf{1}) = 10 \times \mathbf{0} = 0$

$10 + \mathbf{11} = 21$

$10 - 11 = 10 - (\mathbf{10} + 1) = 10 - 10 + \mathbf{1} = 0 + (\mathbf{-1}) = -1$

$11 - 10 = (10 + 1) - (10 + 0) = 10 \times (1 - 1) + 1 \times (1 - \mathbf{0}) = 0 + 1 = 1$

$10 + \mathbf{19} = 29$

$10 - 19 = 10 - (10 + \mathbf{9}) = (10 - \mathbf{0}) + (0 - \mathbf{19}) = 0 + \mathbf{(-9)} = -9$

$19 - 10 = (10 + \mathbf{9}) - 10 = (10 - \mathbf{10}) + (\mathbf{9} - 0) = 0 + \mathbf{9} = 9$

$14 + \mathbf{4} = 18$

$14 - 4 = \mathbf{10} \times (1 - 0) + 1 \times (4 - 4) = 10 \times 1 + 1 \times \mathbf{0} = 10 + \mathbf{0} = 10$

$4 - 14 = 10 \times (0 - \mathbf{1}) + 1 \times (4 - 4) = 10 \times (-1) + 1 \times \mathbf{0} = 10 \times (-1) + \mathbf{0} = \mathbf{-10}$

$16 + 9 = 16 + (\mathbf{4} + 5) = (16 + \mathbf{4}) + 5 = ((10 + \mathbf{6}) + 4) + 5 = (10 + (\mathbf{6} + 4)) + 5$

$= (10 + \mathbf{10}) + 5 = 10 \times \mathbf{2} + 5 = 20 + \mathbf{5}$

$16 + 9 = (10 + \mathbf{6}) + 9 = (10 + (5 + \mathbf{1})) + 9$

$= ((10 + 5) + \mathbf{1}) + 9 = (10 + \mathbf{5}) + (\mathbf{1} + 9)$

$= (10 + \mathbf{5}) + 10 = 10 + \mathbf{5} + 10 = 10 + \mathbf{10} + 5 = 10 \times \mathbf{2} + 5 = 20 + 5 = 25$

$16 - 9 = 10 \times (\mathbf{1} - 0) + 1 \times (6 - \mathbf{9}) = 10 \times 1 + \mathbf{(-3)} = 10 + \mathbf{(-3)} = 7$

$9 - 16$

$= 10 \times (0 - \mathbf{1}) + 1 \times (9 - \mathbf{6})$

$= 10 \times \mathbf{(-1)} + 1 \times \mathbf{3}$

$= \mathbf{-10} + \mathbf{3}$

$= 3 + \mathbf{(-10)}$

$= 3 - \mathbf{10}$

$= 1 \times (3 - \mathbf{10})$

$= 1 \times \mathbf{(-7)}$

$= -7$

-10 + 3

$= -(7 + 3) + 3 = 3 + (-(7 + 3)) = 3 + (-(3 + 7)) = 3 - (3 + 7)$

$= 3 - 3 - 7 = 0 - 7 = -7$

39 + 58

$= (30 + 9) + (50 + 8)$

$= 30 + 50 + 9 + 8$

$= 30 + 50 + 7 + (2 + 8)$

$= 30 + 50 + 7 + 10$

$= 90 + 7$

$= 97$

$30 + 50 + 7 + 10 = 10 \times (3 + 5 + 1) + 7 = 10 \times (8 + 1) + 7$

$= 10 \times 9 + 7 = 90 + 7 = 97$

$39 - 58 = 10 \times (3 - 5) + 1 \times (9 - 8) = 10 \times (-2) + 1 \times 1$

$= 10 \times (-1) + ((-1) \times 10 + 1)$

$= -10 + (1 + (-1) \times 10) = -10 + (1 + (-1) \times (1 + 9)) = -10 + (1 + ((-1) + (-9)))$

$= -10 + (1 + (-1)) + (-9) = -10 + (1 - 1) + (-9) = -10 + 0 + (-9) = -10 + (-9)$

$= (-10) \times 1 + (-1) \times 9 = -19$

$78 + 85 + 69 = (70 + 80 + 60) + (8 + 5 + 9)$

$= (70 + 80 + 60) + (8 + (2 + \mathbf{3}) + 9)$

$= (70 + 80 + 60) + ((\mathbf{10} + 3) + 9)$

$= (70 + 80 + 60) + (\mathbf{10} + (3 + 9))$

$= (70 + 80 + 60) + (\mathbf{10} + ((2 + 1) + 9))$

$= (70 + 80 + 60) + (\mathbf{10} + (2 + 10))$

$= (70 + 80 + 60) + (\mathbf{10} \times 2 + 2)$

$= 10 \times (7 + 8 + 6 + 2) + 1 \times \mathbf{2}$

$= 10 \times (15 + 6 + 2) + 1 \times \mathbf{2}$

$= 10 \times (21 + 2) + \mathbf{1} \times 2$

$= 10 \times 23 + 1 \times \mathbf{2}$

$= 10 \times (10 \times \mathbf{2} + 3) + 1 \times 2$

$= 100 \times \mathbf{2} + 10 \times \mathbf{3} + 1 \times 2 = 200 + 30 + 2 = 232$

$98 - 24 - 58 = (98 - 24) - 58 = 74 - 58 = 16$

$\mathbf{98 - 24 - 58} = 98 - (\mathbf{24} + 58) = 98 - \mathbf{82} = 16$

$34 - 45 - 76 + 27 = 34 - (45 + \mathbf{76}) + 27 = 34 + 27 - (45 + \mathbf{76}) = 61 - \mathbf{121}$

$= (60 + 1) + ((-100) + (-20) + (-1))$

$= -100 + (\mathbf{60} + (-20)) + (1 + (-1))$

$= -100 + \mathbf{40} + 0$

$= -100 \times 1 + \mathbf{10} \times 4 + 1 \times 0$

$= -100 \times 0 + (\mathbf{10} \times 4 + (-10) \times 10) + 0$

$= 0 + 10 \times 4 + 10 \times (\mathbf{-10}) + 0 = 10 \times (4 + (\mathbf{-10})) = 10 \times (\mathbf{-6}) = (\mathbf{-6}) \times 10 = -60$

$= 34 + 27 + \textbf{(-45)} - 76 = 61 + (-(45 + 76)) = 61 + \textbf{(-121)} = 61 - \textbf{(61} + 60)$

$= 61 + \textbf{(-61)} - 60 = \textbf{-60}$

$0 \times 0 = \textbf{0},\ 0 \times 1 = \textbf{0},\ 0 \times (-1) = \textbf{0},\ 1 \times 0 = \textbf{0},$ and $(-1) \times 0 = \textbf{0}$

$3 \times 2 = 3 + \textbf{3} = 6,$ and $(-3) \times 2 = -3 + (-3) = -6$

$3 \times (-2) = 3 \times (-1) \times \textbf{2} = \textbf{(-1)} \times 3 \times 2 = -1 \times \textbf{6} = \textbf{-6}$

$-3 \times (-2) = -1 \times 3 \times \textbf{(-1)} \times 2 = -1 \times \textbf{(-1)} \times 3 \times 2 = \textbf{6}$

$12 \times \textbf{3} = 12 + 12 + 12 = 36 = (10 + \textbf{2}) \times 3 = 10 \times 3 + \textbf{2} \times 3 = 30 + \textbf{6} = \textbf{36}$

15×12

$= (10 + \textbf{5}) \times 12$

$= 10 \times 12 + \textbf{5} \times 12$

$= 10 \times (10 + \textbf{2}) + 5 \times (10 + \textbf{2})$

$= 10 \times \textbf{10} + 10 \times 2 + 5 \times \textbf{10} + 5 \times 2$

$= \textbf{100} + 20 + 50 + 10$

$= \textbf{180}$

$0.1 + \textbf{0.1} = 0.2$ $\qquad\qquad\qquad\qquad$ $0.1 - \textbf{0.1} = 0$

$4.4 + 32.7$

$= (1 \times 4 + 0.1 \times \textbf{4}) + (10 \times \textbf{3} + 1 \times \textbf{2} + 0.1 \times 7)$

$= 10 \times (0 + \textbf{3}) + 1 \times (4 + \textbf{2}) + 0.1 \times (4 + 7)$

$= 10 \times 3 + 1 \times (4 + \textbf{2}) + 0.1 \times (\textbf{10} + 1)$

$= 10 \times 3 + 1 \times (4 + \textbf{2} + 1) + 0.1 \times 1 = 30 + \textbf{7} + 0.1 = \textbf{37.1}$

$4.4 - 32.7$

$= (1 \times \mathbf{4} + 0.1 \times 4) - (10 \times \mathbf{3} + 1 \times 2 + 0.1 \times \mathbf{7})$

$= 10 \times (0 + \mathbf{(-3)}) + 1 \times (4 + (-2)) + 0.1 \times (\mathbf{4} + (-7))$

$= \mathbf{10} \times (-3) + 1 \times 2 + 0.1 \times (-3)$

$= 10 \times ((-2) + (-1)) + 1 \times \mathbf{2} + 0.1 \times (-3)$

$= 10 \times (-2) + \mathbf{10} \times (-1) + 1 \times \mathbf{2} + 0.1 \times (-3)$

$= 10 \times (-2) + (-10) + 1 \times \mathbf{2} + 0.1 \times (-3)$

$= 10 \times (-2) + 1 \times \mathbf{(-10)} + 1 \times \mathbf{2} + 0.1 \times (-3)$

$= 10 \times (-2) + 1 \times (\mathbf{(-10)} + 2) + 0.1 \times (-3)$

$= 10 \times \mathbf{(-2)} + 1 \times (-8) + 0.1 \times (-3) = \mathbf{-28.3}$

$1.8 \times \mathbf{2}$

$= \mathbf{1.8} + \mathbf{1.8} = 1 \times (1 + \mathbf{1}) + 0.1 \times (\mathbf{8} + 8) = 1 \times (1 + \mathbf{1}) + 0.1 \times (\mathbf{10} + 6)$

$= 1 \times (1 + \mathbf{1} + 1) + 0.1 \times \mathbf{6} = 1 \times \mathbf{3} + 0.1 \times \mathbf{6} = \mathbf{3.6}$

$\mathbf{1.8 + 1.8}$

$= (1 + 0.8) \times \mathbf{2} = 1 \times 2 + 0.8 \times 2$

$= 1 \times 2 + 0.1 \times \mathbf{8} \times 2$

$= 1 \times 2 + 0.1 \times \mathbf{16}$

$= 1 \times 2 + 0.1 \times (\mathbf{10} + 6)$

$= 1 \times 2 + 0.1 \times 10 + 0.1 \times \mathbf{6}$

$= 1 \times (\mathbf{2} + 1) + 0.1 \times 6$

$= 1 \times \mathbf{3} + 0.1 \times \mathbf{6} = \mathbf{3.6}$

$0.3 \times 0.2 = 0.1 \times 3 \times 0.1 \times \mathbf{2} = 0.1 \times 0.1 \times 3 \times \mathbf{2} = 0.01 \times \mathbf{6} = \mathbf{0.06}$

0.3 x 0.8

= 0.1 x **3** x 0.1 x 8

= 0.1 x 0.1 x **3** x 8

= 0.01 x **24**

= 0.01 x (**20** + 4)

= 0.01 x 20 + 0.01 x **4**

= 0.01 x **2** x 10 + 0.01 x **4**

= 0.01 x 10 x **2** + 0.01 x 4

= 0.1 x **2** + 0.01 x 4

= **0.2** + 0.04

= **0.24**

1.9 x 0.7

= (1 + 0.9) x 0.7

= 1 x **0.7** + 0.9 x 0.7

= 0.7 + 0.1 x **0.1** x 9 x 7

= 0.7 + 0.01 x (60 + **3**)

= 0.7 + 0.01 x (10 x **6** + 3)

= 0.7 + (0.01 x 10) x **6** + 0.03

= 0.7 + 0.1 x **6** + 0.03

= 0.7 + **0.6** + 0.03

= 0.1 x (**7** + 6) + 0.03

= 0.1 x (**10** + 3) + 0.03

= 0.1 x 10 + **0.3** + 0.03

= **1** + 0.3 + 0.03 = **1.33**

5.9 x 9.8

= (5 + **0.9**) x (**9** + 0.8)

= 5 x (**9** + 0.8) + 0.9 x (**9** + 0.8)

= 5 x 9 + **5** x 0.8 + 0.9 x 9 + 0.9 x 0.8

= 45 + 5 x (0.1 x **8**) + 0.1 x (9 x **9**) + 0.1 x 0.1 x (9 x **8**)

= 45 + 5 x (8 x 0.1) + 0.1 x 81 + 0.01 x **72**

= 45 + (5 x **8**) x 0.1 + 0.1 x 81 + 0.01 x **72**

= 45 + 40 x 0.1 + 0.1 x 81 + 0.01 x **72**

= 45 + 0.1 x 40 + 0.1 x 81 + 0.01 x **72**

= 45 + 0.1 x (40 + **81**) + 0.01 x **72**

= 45 + 0.1 x 121 + 0.01 x 72

= 40 + 5 + 0.1 x (100 + **20** + 1) + 0.01 x (70 + 2)

= 40 + 5 + 0.1 x 100 + 0.1 x **20** + 0.1 + 0.01 x **70** + 0.01 x 2

= 40 + 5 + 0.1 x 10 x **10** + 0.1 x 10 x 2 + 0.1 + 0.01 x 10 x **7** + 0.01 x 2

= 40 + 5 + **10** + 2 + 0.1 + 0.1 x 7 + 0.01 x 2

= 50 + 7 + 0.1 x (1 + **7**) + 0.01 x 2

= 50 + 7 + 0.1 x **8** + 0.01 x 2

= **57.82**

0.04 x 7.9

= 0.04 x (7 + 0.9)

= 0.04 x 7 + **0.04** x 0.9

= (0.01 x **4**) x 7 + (0.01 x **4**) x 0.9

= 0.01 x (**4** x 7) + (0.01 x 4) x (**0.1** x 9)

= 0.01 x (**4** x 7) + 0.01 x **0.1** x 4 x 9

= 0.01 x **28** + 0.001 x **36**

= 0.01 x **28** + 0.001 x (30 + **6**)

= 0.01 x **28** + 0.001 x (10 x **3** + **6**)

= 0.01 x **28** + 0.01 x **3** + 0.001 x **6**

= 0.01 x (**28** + **3**) + 0.001 x **6**

= 0.01 x **31** + 0.006 = 0.01 x (30 + **1**) + 0.006

= 0.01 x (10 x **3**) + 0.01 x **1** + 0.006

= **0.1** x 3 + 0.01 x **1** + 0.006

= **0.3** + 0.01 + 0.006 = **0.316**

1 / **1** = 1 since 1 = 1 x 1, and -1 / 1 = **-1** since -1 = 1 x (-1) = (-1) x 1

1 / (-1) = **-1** since 1 = (-1) x (-1) = 1

-1 / (-1) = **1** since -1 = (-1) x 1 = -1, and 2 / 1 = 2 since 2 = 1 x 2

-2 / 1 = **-2** since -2 = 1 x (-2) = (-2) x 1

2 / (-1) = **-2** since 2 = (-1) x (-2) = (-1) x (**(-1)** x 2)

= ((-1) x (**-1**)) x 2 = 1 x 2 = 2

2 / 2 = 1 since 2 = 2 x 1, and -2 / 2 = **-1** since -2 = 2 x (-1) = (-1) x 2 = -2

2 / -2 = **-1** since 2 = (-2) x (-1) = (-1) x 2 x (-1) = (-1) x (-1) x 2 = 1 x 2

4 / 2 = **2** since 4 = 2 x **2**, and 6 / **3** = 2 since 6 = **3** x 2

10 / **1** = 10 since 10 = 1 x **10**, and 12 / **3** = 4 since 12 = **3** x 4

$28 / 2 = (20 + 8) / 2 = (20 / 2) + (8 / 2) = 10 + 4 = 14$

$= (30 - 2) / 2 = (30 / 2) - (2 / 2) = 15 - 1 = 14$

$1248 / 4$

$= (1200 + 40 + 8) / 4$

$= (1200 / 4) + (40 / 4) + (8 / 4)$

$= ((100 \times 12) / 4) + ((10 \times 4) / 4) + ((1 \times 8) / 4)$

$= (100 \times (12 / 4)) + (10 \times (4 / 4)) + (1 \times (8 / 4))$

$= 100 \times 3 + 10 \times 1 + 1 \times 2$

$= 300 + 10 + 2$

$= 312$

$1248 / 4 / 2$

$= (1248 / 4) / 2$

$= 1248 / (4 \times 2)$

$= 1248/8$

$= 156$

$1248 / 4 / 2$

$= 312 / 2$

$= (100 \times 3 + 10 \times 1 + 1 \times 2) / 2$

$= 100 \times (3 / 2) + 10 \times (1 / 2) + 1 \times (2 / 2)$

$= 100 \times (1 + 1 / 2) + 10 \times (1 / 2) + 1 \times (2 / 2)$

$= 100 \times 1 + 100 / 2 + 10 / 2 + 1 \times 1$

$= 100 \times 1 + (10 \times \mathbf{10}) / 2 + (1 \times \mathbf{10}) / 2 + 1 \times 1$

$= 100 \times 1 + 10 \times (\mathbf{10} / 2) + 1 \times (\mathbf{10} / 2) + 1 \times 1$

$= 100 \times 1 + 10 \times \mathbf{5} + 1 \times \mathbf{5} + 1 \times 1$

$= 100 \times 1 + 10 \times \mathbf{5} + 1 \times (\mathbf{5} + 1)$

$= 100 \times 1 + 10 \times \mathbf{5} + 1 \times \mathbf{6}$

$= \mathbf{156}$

$1 / 2 = (0.1 \times \mathbf{10}) / 2 = 0.1 \times (\mathbf{10} / 2) = 0.1 \times \mathbf{5} = \mathbf{0.5}$

$= 0.5$ since $0.5 \times \mathbf{2} = 1$

$0.1 / 2 = (0.01 \times \mathbf{10}) / 2 = 0.01 \times (\mathbf{10} / 2) = 0.01 \times \mathbf{5} = \mathbf{0.05}$

$3 / 2 = (2 + \mathbf{1}) / 2 = 2 / 2 + \mathbf{1} / 2 = 1 + \mathbf{0.5} = \mathbf{1.5}$

$15 / 6$

$= (12 + \mathbf{3}) / 6$

$= 12 / 6 + \mathbf{3} / 6$

$= 2 + (1 \times \mathbf{3}) / 6$

$= 2 + ((0.1 \times \mathbf{10}) \times 3) / 6$

$= 2 + (0.1 \times (\mathbf{10} \times 3)) / 6$

$= 2 + (0.1 \times \mathbf{30}) / 6$

$= 2 + 0.1 \times (\mathbf{30} / 6)$

$= 2 + 0.1 \times \mathbf{5}$

$= 2 + \mathbf{0.5}$

$= \mathbf{2.5}$

3.9 / 2

$= (3 + \textbf{0.9}) / 2$

$= (1 \times 3 + 0.1 \times \textbf{9}) / 2$

$= (1 \times 3) / 2 + (0.1 \times \textbf{9}) / 2$

$= 1 \times (\textbf{3} / 2) + 0.1 \times (\textbf{9} / 2)$

$= 1 \times ((\textbf{2} + 1) / 2) + 0.1 \times ((\textbf{8} + 1) / 2)$

$= 1 \times ((2 / 2) + (1 / 2)) + 0.1 \times ((\textbf{8} / 2) + (1 / 2))$

$= 1 \times (1 + \textbf{0.5}) + 0.1 \times (\textbf{4} + 0.5)$

$= 1 \times 1 + 0.5 + 0.4 + 0.05$

$= 1 \times 1 + 0.1 \times (\textbf{5} + 4) + 0.01 \times \textbf{5}$

$= 1 \times 1 + 0.1 \times \textbf{9} + 0.01 \times \textbf{5}$

$= \textbf{1.95}$

2 / 4

$= (1 \times \textbf{2}) / 4$

$= ((0.1 \times \textbf{10}) \times 2) / 4$

$= (0.1 \times (\textbf{10} \times 2)) / 4$

$= (0.1 \times \textbf{20}) / 4$

$= 0.1 \times (\textbf{20} / 4)$

$= 0.1 \times \textbf{5}$

$= \textbf{0.5}$

$1 / 4 = (0.1 \times \textbf{10}) / 4 = 0.1 \times (\textbf{10} / 4) = 0.1 \times ((\textbf{8} + 2) / 4)$

$= 0.1 \times ((8 / 4) + (2 / 4)) = 0.1 \times (2 + \textbf{0.5}) = 0.1 \times 2 + \textbf{0.05}$

$= 0.1 \times 2 + \textbf{0.01} \times 5 = \textbf{0.25}$

9.7 / 4

= (9 + **0.7**) / 4

= 9 / 4 + **0.7** / 4

= (8 + 1) / 4 + (0.1 x **7**) / 4

= 2 + 1 / 4 + 0.1 x (**7** / 4)

= 2 + 0.25 + 0.1 x ((**4** + 3) / 4)

= 2 + 0.25 + 0.1 x (**1** + 3 / 4)

= 2 + 0.25 + 0.1 x (1 + (0.1 x **30**) / 4)

= 2 + 0.25 + 0.1 x (1 + 0.1 x (**30** / 4))

= 2 + 0.25 + 0.1 x (1 + 0.1 x ((**28** + 2) / 4))

= 2 + 0.25 + 0.1 x (1 + 0.1 x (**7** + 2 / 4))

= 2 + 0.25 + 0.1 x (1 + 0.1 x (**7** + 0.5))

= 2 + 0.25 + 0.1 x (1 + **0.7** + 0.05)

= 2 + 0.25 + 0.1 + **0.07** + 0.005

= 2 + 0.2 + **0.05** + 0.1 + 0.07 + 0.005

= 2 + 0.1 x (**2** + 1) + 0.01 x (5 + 7) + 0.005

= 2 + 0.1 x **3** + 0.01 x (10 + 2) + 0.005

= 2 + 0.1 x **3** + 0.01 x 10 + 0.01 x 2 + 0.005

= 2 + 0.1 x **3** + 0.1 x 1 + 0.01 x 2 + 0.005

= 2 + 0.1 x **4** + 0.01 x 2 + 0.005

= **2.425**

1 / 0.1 = (0.1 x **10**) / 0.1 = (10 x **0.1**)/ 0.1 = 10 x (**0.1** / 0.1) = 10 x 1

1 / 0.2

$= (0.1 \times \mathbf{10}) / 0.2$

$= (10 \times \mathbf{0.1}) / 0.2$

$= 10 \times (\mathbf{0.1} / 0.2)$

$= 10 \times ((0.01 \times \mathbf{10}) / 0.2)$

$= 10 \times (0.01 \times (\mathbf{10} / 0.2))$

$= 10 \times (0.01 \times ((0.1 \times \mathbf{100}) / 0.2))$

$= 10 \times (0.01 \times ((0.1 \times (2 \times \mathbf{50})) / 0.2))$

$= 10 \times (0.01 \times (((0.1 \times 2) \times \mathbf{50}) / 0.2))$

$= 10 \times (0.01 \times ((0.2 \times \mathbf{50}) / 0.2))$

$= 10 \times (0.01 \times ((\mathbf{50} \times 0.2) / 0.2))$

$= 10 \times (0.01 \times (\mathbf{50} \times (0.2 / 0.2)))$

$= 10 \times (0.01 \times (\mathbf{50} \times 1))$

$= 10 \times (0.01 \times (10 \times \mathbf{5}))$

$= 10 \times (0.1 \times \mathbf{5})$

$= (10 \times 0.1) \times \mathbf{5}$

$= 1 \times \mathbf{5} = \mathbf{5}$

2 / 0.4

$= (0.1 \times \mathbf{20}) / 0.4$

$= (0.1 \times (4 \times \mathbf{5})) / 0.4$

$= ((0.1 \times 4) \times \mathbf{5}) / 0.4$

$= (0.4 \times \mathbf{5}) / 0.4$

$= (\mathbf{5} \times 0.4) / 0.4$

$= \mathbf{5} \times (0.4 / 0.4) = \mathbf{5} \times 1 = \mathbf{5}$

7.9 / 0.4

= (7 + 0.9) / 0.4

= 7 / **0.4** + 0.9 / **0.4**

= (4 + **3**) / 0.4 + (0.1 x **9**) / 0.4

= 4 / 0.4 + 3 / 0.4 + 0.1 x (9 / 0.4)

= 10 + (1 x **3**) / 0.4 + 0.1 x (1 x **9**) / 0.4

= 10 + ((0.1 x **10**) x 3) / 0.4 + 0.1 x 0.1 x **10** x 9 / 0.4

= 10 + (0.1 x **30**) / 0.4 + 0.1 x (0.1 x **90**) / 0.4

= 10 + 0.1 x ((**28** + 2) / 0.4) + 0.1 x 0.1 x (**88** + 2) / 0.4

= 10 + 0.1 x (**28** / 0.4 + 5) + 0.1 x 0.1 x (**88** / 0.4 + 5)

= 10 + 0.1 x ((**7** x 4) / 0.4 + 5) + 0.1 x 0.1 x (22 x 4 / 0.4 + 5)

= 10 + 0.1 x (**7** x (4 / 0.4) + 5) + 0.1 x 0.1 x (22 x (4 / 0.4) + 5)

= 10 + 0.1 x (**7** x 10 + 5) + 0.1 x 0.1 x (22 x 10 + 5)

= 10 + 0.1 x **7** x 10 + 0.1 x **5** + 0.1 x (0.1 x (22 x 10) + 0.1 x 5)

= 10 + 7 + **0.5** + 0.1 x (22 + 0.5)

= 10 + 7 + **0.5** + 0.1 x (20 + **2**)+ 0.05

= 10 + 7 + **0.5** + 0.1 x (2 x 10) + 0.2 + 0.05

= 10 + 7 + **0.5** + 2 + 0.2 + 0.05

= 10 + 9 + **0.7** + 0.05

= **19.75**

Note again, you don't have to do calculations the way above.

Each calculation just shows that **there are many ways we can calculate**.

And there can be many ways you can break a particular value into many other values, and put them together to get the same particular value as many ways as you can think of.

And the same is true for expressions of many kinds, too.
For instance, we can get:

$$2a + b = a + a + b = 2a + 2b - b = 2(a + b) - b = \ldots$$

$$a^2 + 2a = a^2 + 2a + 1 - 1 = (a + 1)^2 - 1 = \ldots$$

$$a^2 + b^2 = a^2 + b^2 + 2ab - 2ab = (a + b)^2 - 2ab = \ldots$$

www.ingramcontent.com/pod-product-compliance
Lightning Source LLC
Chambersburg PA
CBHW081446170526
45166CB00008B/2327